Accelerating Test, Validation and Debug of High Speed Serial Interfaces

Yongquan Fan · Zeljko Zilic

Accelerating Test, Validation and Debug of High Speed Serial Interfaces

Springer

Dr. Yongquan Fan
High Performance Analog
Texas Instruments
12500 TI Blvd, Dallas, TX 75243
USA
yfan@ti.com

Prof. Zeljko Zilic
Department of Electrical & Computer Engineering
McGill University
University Street 3480
H3A 2A3 Montréal, Québec
Canada
zeljko.zilic@mcgill.ca

ISBN 978-90-481-9397-4 e-ISBN 978-90-481-9398-1
DOI 10.1007/978-90-481-9398-1
Springer Dordrecht Heidelberg London New York

Library of Congress Control Number: 2010938288

Printed on acid-free paper

Springer is part of Springer Science+Business Media (www.springer.com)

To Ji Lei

Yongquan

To Kasia, Maria, Ivan Alexander and Pauline Veronica

Zeljko

Acknowledgments

Authors would like to thank our colleagues and friends at McGill University, Agere Systems/LSI Corporation and Texas Instruments for providing a good environment to conduct this research and finish the book. We especially thank Dr. Yi Cai at LSI for providing longstanding and wise technical advice and co-authoring two papers, Professor Gordon Roberts at McGill for giving research guidelines, Bill Kempler and Sam Yingsheng Tung at Texas Instruments for reviewing the manuscript. A part of the work leading to this book was undertaken while Yongquan Fan was pursuing first his M. Eng. and then Ph. D. degree at McGill University.

Our thanks also go to Liming Fang, Anant Verma, Bill Burchanowski, and Sandeep Kumar for collaborating on a work that led to co-authoring one paper. We also appreciate the technical support and help from the whole PHY and Storage team at Agere/LSI, especially from Angshu Bhattacharyya, Joe Martone, John Janney, Suri Basharapandiyan, Bernhard Laschinsky, Kevin Richter, Paul Hua, Tom Gibson, Kahn Neguen, Bob Hain and Ken Paist.

In addition, we thank the many people who have given valuable advice, feed back and help at conferences and various other occasions. Special thanks are extended to Mohamed Hafed at DFT Microsystems, Takahiro J. Yamaguchi at Advantest Corporation, Yang Liang at Maxim, Steve Sunter at Logic Vision, Mike Li at Wavecrest (now at Altera), Xu Fang at Teradyne, Joe Venable, Shu Xia and Wai Lee at Texas Instruments, Jwo Cheng and Carlo DiGiovanni from GigOptix, Minh Tran from Teledyne, Mani Soma at University of Washington, Luo He at Concordia University, Warren Gross, Kasia Radecka, Atanu Chattopadhyay, Man-Wah Chiang, Rong Zhang, Milos Prokic and Jean-Samuel Chenard at McGill University

Altera Corporation has provided also a great support throughout the years, from their devices and intellectual property, to the encouragement and research funding. Special thanks go to Steve Brown, Fort Blair, Tomasz Czajkowski and David Mendel.

At Springer Science+Business Media, Mark de Jongh and Cindy Zitter were always there for us, providing encouragement and cheering the race to the finish line.

As with the recent book (Generating Hardware Assertion Checkers, Springer 2008) co-written by Marc Boulé and Zeljko Zilic, the cover page art was drawn by the second author's children. This time, Maria Zilic took artistic leadership, with

Ivan and Pauline Zilic doing their share to together depict the effects of the jittery clock in an appealing and a surprisingly lucid way.

Last but not least, we would like to thank our whole families for their love and support over the years. Without their support, it would be impossible to undertake all the work and finally write the book.

Table of Contents

1 Introduction

Abstract *This chapter brings out the motivation for our research, together with the rudimentary background information on high-speed serial interface standards. The chapter then enumerates the challenges that we are facing in high-speed serial interface testing, validation and debugging, and finally outlines our solutions to some of the most pressing challenges.*

1.1 Motivation

The High Speed Serial Interface (HSSI), which is interchangeably referred to as Serializer/Deserializer (SerDes) or simply a Transceiver, is a cornerstone of modern communication, from high-performance communication and computation infrastructure, to the desktop computing and increasingly even to consumer electronics.

As the HSSI data rate reaches a few Giga-bits per second (Gbps) and continues to increase, the room for its timing deviation, i.e., *jitter*, is getting tighter and tighter. To achieve high data rates, sophisticated techniques such as equalization and pre-compensation have now become common in HSSIs. With the concurrent increase in design complexity and decrease in the timing budget, the traditional "Guaranteed by Design" paradigm is not valid anymore. It is hence becoming imperative to qualify the tight timing and other specifications in silicon in order to guarantee the design quality.

The post-silicon qualification usually consists of three processes: validation of the first set of fabricated devices, characterization of the devices under all settings across Process, Voltage and Temperature (PVT) corners, and production testing. *Validation* emphasizes on verifying complete device functionality, including parameter values and electrical characteristics.

Because of the increasing design complexity, close to 25% of all design resources at Intel, for example, are now spent on post-silicon validation [20]. Validation is usually performed in a lab environment, where standard instruments, such as oscilloscopes, signal generators and logic analyzers are used. The standard measurement equipment is also referred to as *bench* equipment, and the validation and testing approaches based on standard equipment are also referred to as bench test solutions.

Characterization is the process that is more concentrated on verifying that the device can work under all settings and can accommodate process variations allowed in manufacturing. Characterization can in principle be done either in the lab or on Automatic Test Equipment (ATE), and there is a significant push to achieve

Y. Fan, Z. Zilic, *Accelerating Test, Validation and Debug of High Speed Serial Interfaces*,
DOI 10.1007/978-90-481-9398-1_1,

as much characterization as possible on an ATE, provided that it is economical to do so, hence the methods that speed up the characterization on production ATE testers are of great interest.

Production testing determines the pass/fail of each device in a mass production environment. The test throughput is paramount in production testing because it directly affects the device cost. ATE is widely used in production because of its high throughput, but at high clock rates associated with the emerging HSSI devices, the ATE equipment might be either too costly or not fast enough.

Among all the HSSI parameters that we need to qualify, Bit Error Rate (BER) and Jitter are the critical ones. *BER* is the bit error probability of the system, which shows how well the system works in an end-to-end manner. *Jitter* is the deviation of a signal from its ideal timing, and it is the major cause of bit errors in a well-designed high-speed communication system. It usually is expressed relative to the clock signal, where such deviations can cause bits to be incorrectly latched. In data communications, we now usually talk about bit errors caused by jitter. Jitter specifications are normally defined at 10^{-12} BER rates or lower.

It is very challenging and costly to qualify the timing specification mainly for three reasons:

- ATE has been widely used in production testing because of its high throughput. However, the increasing demand for more bandwidth is continuously pushing the data communication rate higher, at a pace faster than the test equipment evolves; systematic HSSI testing solutions on ATE for data rates above 6 Gbps are not commercially mature yet [8].
- Cost goals set by the marketplace demand competitive test solutions – testing needs to be done as fast as possible, while using as inexpensive equipment as possible; it is infeasible to use traditional lab instruments in a production environment because it takes hours or even days to qualify the jitter and BER performance.
- Validating the jitter performance across PVT corners is becoming necessary with the continuing scale of the process technology, but the validation is very time-consuming; shortening the validation time (including debugging when necessary) would directly reduce the time-to-market, which provides great competitive advantages in gaining profit and market share.

Motivated by the great economic significance in qualifying HSSIs, this research concentrates on developing HSSI test and characterization methodologies to address the above challenges. We aim to qualify HSSIs accurately and cost-effectively, yet overcoming ATE limitations.

1.1.1 HSSI Technology Trends

With the evolution of information technology during the past few decades, commodities such as cell phones and computers have become commonplace. They have made it a reality for people all over the world to share information or communicate directly in one way or another. This evolution has caused a drastic increase in the amount of information generated and the number of end users that need to access the information. As a platform to communicate information, Internet has become the main driver for technology innovation and bandwidth growth. The key to meeting the increasing demand for bandwidth is the HSSI.

Figure 1-1 illustrates the structure of an Ethernet-based communication infrastructure. In this network, HSSIs are widely adopted into backplane applications, short and long-haul communications, mass storage access networking, and computer peripherals. The bandwidth requirement of the HSSIs depends on the proximity to the end-user and the location in the network.

In recent years, different serial communication protocols have arisen to address a wide set of communication applications, such as Ethernet, XAUI, GPON and SATA.

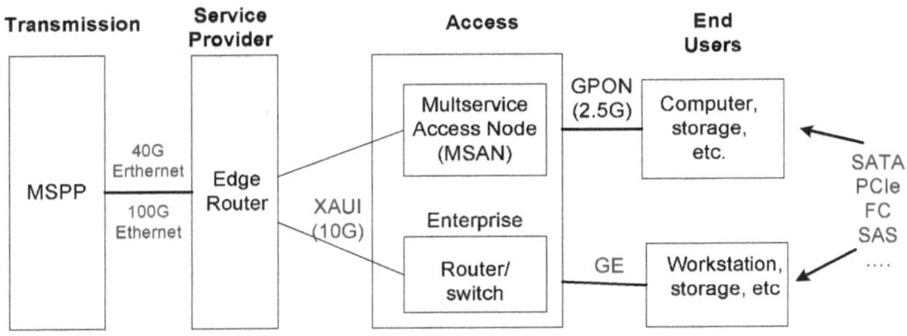

Fig. 1-1. HSSIs in communication infrastructure

The HSSI protocols are continuously evolving to higher speeds to meet the demand for higher bandwidth. One example is the Serial ATA (SATA): when the SATA 1.0 Working Group was formed in February 2000 to design SATA for desktops, the target speed was only 1.5Gbps; in 2004 it evolved to 3Gbps and now SATA 3.0 provides 6Gbps data rate [22]. Another example is the Ethernet: 10Gbps Ethernet (10GbE) was first published by IEEE in 2002 [9]. Currently it is the fastest matured standard, but IEEE is already in development of 40GbE and 100GbE.

The increasing bandwidth requirements are driving silicon vendors to provide HSSIs with higher speeds. In 2002, the highest data rate in Altera Field Programmable Gate Arrays (FPGAs) was only 1.25 Gbps per channel, available in its Mercury devices [10]; now in Altera Stratix IV GT FPGAs, the rate has increased to 11.3 Gbps per channel, with up to 48 transceivers each device [11]. Another major FPGA provider, Xilinx, provides up to thirty-six 11.2 Gbps transceivers in its Virtex-6 and Spartan-6 FPGAs, capable of supporting 40G/100G applications [12].

The increase in data rate and integration in FPGAs is just a snapshot of the trends in the whole semiconductor industry. Moore's law is still driving the industry to double the number of transistors in an integrated device every two years, making it possible to integrate more functions and at the same time provide higher performance.

The aggressive scaling in deep submicron technologies has enabled the System-on-Chip (SoC) integration of a microcontroller/DSP, ADC/DAC, memory blocks, power management, PLL and external interfaces. Many Giga-Hertz serial interfaces are built in SoC type devices with CMOS process. The data rate of the interfaces also scales accordingly. Besides the FPGA providers, some other semiconductor companies have developed HSSIs with data rates up to 10Gbps per channel, and with up to 100 channels per device [13], [14]. It has been a trend to put more and faster HSSIs in a single device to meet the increasing bandwidth demand.

When we keep pushing the speed envelope and increase the integration, many signal integrity related issues arise, such as timing jitter, noise and frequency loss. A few key technologies have been developed recently to address these issues [15]. Most notably, the pre-emphasis and equalization techniques are used to compensate frequency-related losses, especially those related to Printed Circuit Board (PCB) design due to the skin effect and dielectric loss [16].

Pre-emphasis is the process employed in the transmitter to boost the high-frequency components of a data signal before it is launched to the transmission medium. Equalization in the receiver acts as a high-pass filter to the data signal when it enters the receiver and re-shapes the signal in order to interpret the received signal correctly.

With the integration increase, there is a trend to implement multiple data rates in a single HSSI to accommodate multiple protocols. This requires the HSSI capable of providing multiple rate clock signals. The Phase Locked Loop (PLL) is widely used for clock generation, where a Voltage-Controlled Oscillator (VCO) is a key component. There are two types of oscillators: Ring Oscillator (RO) and LC tank oscillator (LC tank). RO has the advantages of small chip area and wide tunable frequency range, but LC tanks provide lower noise and better jitter performance [17], [18].

Multiple data rates can be implemented by changing divider ratios inside the PLL or by providing additional VCOs. In Altera Stratix IV GT FPGAs, the RO can support data rates from 600Mbps to 10.3 Gbps; two LC tanks are also implemented in this device, one with 4.9~6.375 Gbps optimized for PCIe/CEI-6 compliance and the other with 9.9~11.3 Gbps optimized for XLAUI/CAUI/CEI-11G compliance [19].

A side effect of implementing multiple data rates is that the jitter performance of the HSSI can vary across its data range. If the same PLL is used at two speeds, such as 6Gbps and 8.5Gbps, one speed can be susceptible to higher jitter because the two speeds are derived from the same VCO that can only be optimized at one speed. If different PLLs are used to support different data rates, the performance at one data rate does not correlate to another data rate. In either case, good performance at a higher data rate does not guarantee better margin at lower data rates because PLL characteristics may be different.

1.1.2 Qualification Challenges

With the increasing data rate and higher degree of integration, the staggering complexity makes it challenging to design fault-free devices. Post-silicon qualifications are critical in guaranteeing the design quality and the device quality. It is challenging and expensive to qualify the HSSI devices, especially the jitter performance – transmitter jitter and receiver jitter tolerance.

Numerous HSSI standards define jitter performance at the 10^{-12} BER level, which requires running at least 10^{13} bits. This requirement fundamentally limits the test speed: for instance, at 3Gbps data rate, it takes around one hour to run so many bits. With some of the emerging applications demanding 10^{-14} BER, direct measurements are even further from being practical.

In addition, many settings in the HSSI may affect its jitter performance. A few examples include the boost control settings in the equalizer, the bandwidth setting in the PLL and the driver strength settings in the transmitter. These settings are quite common in today's HSSIs. Choosing the optimal setting for the whole HSSI from hundreds or even thousands of available settings is challenging. It requires a tremendous amount of resources in validation and test in order to guarantee the device quality.

Because of the long test time, traditionally the jitter performance of multi-gigabit HSSI devices is only evaluated on bench in limited combinations of PVT. Besides the long test time, another reason that jitter is not qualified in production is the limited availability of ATE instruments, especially for very high-speed applications. For example, to evaluate 8.5GHz FC devices, we prefer the signal generator for the receiver and the digitizer for the transmitter with a bandwidth much higher than 8.5G (such as 15GHz), but they are not commercially mature yet on ATE [8]. Furthermore, there are currently no systematic ATE solutions that can perform complete HSSI testing accurately and cost-efficiently. Most companies only do loopback tests in production to check the functionality. Some HSSI parameters, such as transmitter jitter and receiver jitter tolerance, are assumed to be guaranteed by design.

Unfortunately, the "Guaranteed by Design" quality paradigm is no longer valid while we keep advancing the semiconductor technology and increasing the data rate, which results in tightening the jitter specifications. The devices can increasingly fail just because they do not comply with the timing specifications. According to the data by Collett International, the timing, mixed-signal interfaces, clocking and crosstalk are among the prime failure reasons, each contributing 18% or more to the failure of the first silicon.

Since HSSIs uniquely personify all such issues, they are therefore critical to achieving the overall system quality. It is hence becoming imperative to develop systematic HSSI compliance testing solutions on ATE to distinguish bad devices from good ones in production. This is the only way to ensure the device quality and to eliminate or reduce customer returns.

Since the failures of the first fabricated silicon are now prevalent due to the inability to find the bugs before the fabrication, the post-silicon debugging has become very critical. Here, the goal is to identify the failing cases and *perform the root cause analysis* that will detect the real source of the bug. We foresee more need for HSSI post-silicon debugging and would want to provide the means to facilitate that in an economical way.

Besides the production testing, ATE is also becoming more and more popular in characterization and validation due to its high throughput. A thorough validation and characterization of a design requires performing measurements on different process materials at different temperature and voltage combinations. All these combinations of parameters may lead to measure the same parameter more than 100 times on one device in order to obtain its characteristics under different conditions.

The traditional bench validation approach can no longer meet the requirement of measuring a large amount of parameters in a short time. ATE-based characterization and validation solutions therefore have to be employed to meet the requirements.

Our aim in this book is to develop systematic HSSI testing solutions on ATE that can measure the HSSI standard parameters and design specifications for validation and characterization purposes. To achieve the best test economy, we assume throughout much of our considerations a simplified test flow in production that aims to qualify the key parameters only.

1.1.3 ATE Perspectives

In general, ATE can provide high throughput and has been widely used in production test. It is also more and more widely use for validation and characterization to shorten the time-to-market. To validate, characterize and test HSSIs on ATE, there are several considerations that need to be addressed.

The first concern is the test cost. In early days, HSSI devices were designed as high-performance and high-margin devices. With the introduction of the low cost CMOS processes, most of the Gigahertz HSSIs are now built in high-volume and low-priced SoCs. The large-scale integration makes it however challenging to design a high performance HSSI block in a very noisy SoC environment. It is then becoming more challenging to provide competitive production test solution because of the test time budget.

For an average high-volume SoC, the normal acceptable test time ranges from a few seconds to ten seconds. The HSSI testing is tied with the testing of all other analog and digital blocks in the SoC. As one of many blocks in an SoC, it is expected that the HSSI block testing can be done within 1~2 seconds. For transmitter jitter testing and the receiver jitter tolerance testing, the test time budget is usually limited to a few tens or hundreds milliseconds because there are hundreds of other parameters to test. Test cost control is one of the biggest challenges in a mass-production environment. It is hence urgent to develop jitter testing techniques that can meet the cost requirement.

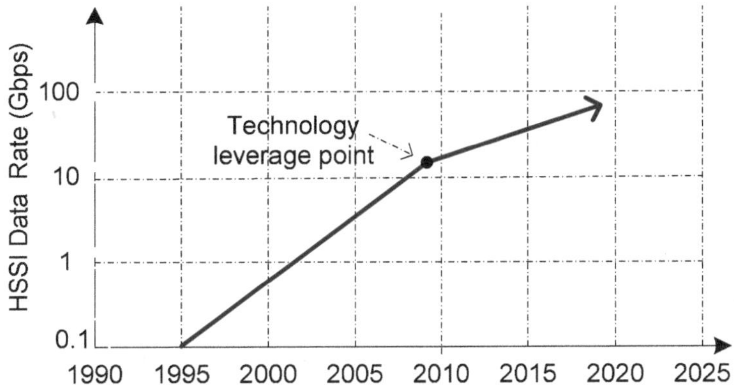

Fig. 1-2. High speed serial interface technology trend

The second major concern is the availability of high-speed instruments on an ATE tester. The increasing bandwidth demand has been pushing the HSSI data rate higher and higher. Figure 1-2 illustrates the historic technology points and future trends [8].

During the past ten years, the data rate has increased from around 1Gbps to the 10 Gbps range. The HSSI I/O frequency is growing faster than the packaging and test fixture technology. Because of the physical limits due to the packaging and test socket, the data rate may slow down somewhat beyond 13 Gbps, shown as the technology leverage point in Figure 1-2 [8]. During the past few years, the ATE industry has made significant progress in providing HSSI test solutions. We note that several ATE suppliers have provided production pin-card solutions up to 6Gbps.

For higher speed, such as 8.5Gbps and 10Gbps applications, systematic ATE production solutions with jitter testing are not commercially mature yet. Normally, only loopback testing is implemented in production to provide limited coverage. It is therefore imperative to address the ATE limitations in order to provide systematic production test solutions for applications above 6Gbps. Another ATE limitation is in multi-lane HSSI testing; most ATE platforms do not have enough high-speed instruments to accommodate the testing of multi-lane HSSI devices.

The third challenge is jitter decomposition and jitter injection. Many modern HSSI standards specify the jitter specifications in term of Deterministic Jitter (DJ) and Random Jitter (RJ), so the procedures for testing, characterization and validation as well need to be cognizant of such types of jitter.

In practice, the traditional concept of *histogram based peak-to-peak jitter* has been replaced the by the concept of Total Jitter (TJ), which is related to a certain BER level. Jitter test solutions need to be capable of decomposing TJ to DJ and RJ. In addition, uncorrelated jitter detection is also needed because it can fail the device in real applications but some jitter measurement techniques cannot detect it.

To conduct a jitter tolerance test for the receiver, we need to have instruments that can deliberately inject controllable amounts of jitter. Depending on applications, we may need to inject PJ, RJ or DJ. Integrated ATE instruments that can inject all these kinds of jitter do not exist currently. It is expected that these issues can be addressed in the test community.

1.2 Contributions

This book addresses the urgent need in the semiconductor industry for cost efficient solutions to qualify HSSI jitter and BER performance [1]. We develop accelerated jitter testing solutions based on existing ATE instrument. We also develop novel low cost, non-ATE solutions that practically overcome the ATE instrument limitation.

We briefly summarize the contributions of this book as the following set of goals that we achieved. We:

- Develop a jitter tolerance extrapolation algorithm that can accelerate jitter tolerance testing by >1000 times [2]. Based on the algorithm, the book proposes a solution for jitter tolerance production testing and a solution for characterization. Using existing ATE instruments, the production testing only takes a few tens milliseconds and the characterization only takes around 1 second, the fastest solution to the best of our knowledge. Direct measurements down to 10^{-12} BER in 3Gbps applications demonstrate the excellent extrapolation accuracy: the discrepancy between the measured results and extrapolation results is within 2 ps [3].
- Present solutions for transmitter jitter testing using the time domain, frequency domain and hybrid approaches based on existing ATE instruments [4]. We manage to achieve sub-picosecond RJ measurement accuracy: the discrepancy between the ATE and bench measurement results is within 0.5ps and the run-to-run variation on ATE is also within 0.5ps. The DJ discrepancy is only a few pico-seconds. The transmitter jitter testing along with other transmitter testing can be done in less than 100 milliseconds while existing solutions usually take a few seconds. In addition, our innovative hybrid approach eliminates some limitations posed by the time domain and the frequency domain approaches, making test results more reliable.
- Propose low-cost HSSI testing solutions without the need for employing high-speed ATE instruments. This is achieved with the following set of sub-goals:

- a) Develop a novel jitter injection technique using the state-of-the-art phase delay lines that can handle clock/data rates of up to 12.5Gbps [5].
- b) Investigate the applications of high-speed relays and propose a versatile loopback-based jitter compliance testing solution [5].
- c) Develop an FPGA-based BER Tester (BERT) for HSSI bit error detection [5].

- Besides exploring the jitter impact to BER, the book also addresses the amplitude noise impact on BER. We investigate the digital Gaussian noise generation and BER testing schemes and propose a novel approach to generate Gaussian noise with excellent tail distribution properties [6].

1.3 Overview of the Book

In the remainder of the book, Chapter 2 presents the background of the research. We first discuss the HSSI technologies and the BER mechanism. BER is a measure of the HSSI overall performance. We then introduce how the timing jitter and the amplitude noise can affect the BER performance.

In Chapter 3, the details of an ATE-based receiver testing solution are presented. We use a high-speed Arbitrary Waveform Generator (AWG) to generate test signals with controllable amounts of injected jitter. Based on the calibrated test signals and the test setup, we develop a jitter tolerance extrapolation algorithm. This algorithm enables us to accelerate the jitter tolerance characterization and production testing by more than 1000 times. Experimental results demonstrate the excellent accuracy of the approach.

In Chapter 4, we present the details of an ATE-based transmitter testing solution. A high bandwidth digitizer is used to capture the transmitter output. We will introduce how the test settings are developed for data acquisition and how the jitter components are extracted. The proposed solution can complete the transmitter testing and characterization in100 milliseconds, and the test accuracy reaches sub-picosecond range.

Chapter 5 discusses the HSSI testing techniques that do not rely on high-speed ATE instruments. We propose a phase-delay line based jitter injection scheme. Based on the novel scheme, an external loopback testing solution is developed to reduce the test cost and also overcome some ATE limitations. By putting the ATE-based approach and the loopback approach together using high-speed relays, we propose a more versatile scheme for HSSI validation, debugging, characterization and production testing.

In Chapter 6, we address BER testing and debugging under noise conditions. Traditionally software is used to simulate/evaluate the BER performance of a communication interfaces under different noise conditions. Even though a software based approach is easy to setup, it is too time consuming to perform low

BER evaluation. Hardware based emulation can greatly speed up the evaluation process. In emulation, a scalable high speed high accuracy Gaussian noise generator is used to emulate noise conditions. In this chapter, we first survey the current hardware based Gaussian noise generation techniques, and then propose a novel approach. Our approach overcomes many limitations of other approaches and is especially suitable for low BER evaluation.

Conclusions are provided in Chapter 7.

2 Background

Abstract *In this chapter, we first introduce the architecture of the HSSI communication interfaces, together with its common applications. Then, the bit error rate (BER) mechanisms are explained, and the jitter phenomenon is dealt with in detail. Introduced are the relevant probabilistic properties and the basics of the simulation and emulation approaches to the modeling of BER effects.*

2.1 High-Speed Serial Communication

The high-speed serial communication interfaces have been widely used in modern communication to deliver fast and robust data transmission. They deliver data at rates from a few Gigabits per seconds to more than 10 Gbps. Traditional single-ended I/O standards, such as in PCI and VME, has by now made way to HSSIs, as they were limited to clock rates of about 200 Mbps due to the combination of noise and loading (fanout) limitations. Differential I/O standards have been used to break the frequency barrier of single-ended I/O standards with common mode rejection. They allow data transmission at higher speeds, although the clock skew issue arises for differential I/O standards when the frequency approaches 1 Gbps and beyond [142].

The technology that enables multiple Gbps high-speed serial communication is often referred to as Clock Data Recovery (CDR). The clock skew concerns are removed by encoding the clock into every data stream, so CDR circuitry provides a mechanism for the clock to track the data. Hence, it eliminates frequency barriers faced by clock synchronous systems. Figure 2-1 illustrates the CDR mechanism: a transmitter embeds the clock in the data stream and a receiver employs specialized CDR circuitry to recover the data, as well as the clock. While this recovery circuit is often used to name the whole communication system as a CDR, in this book, we prefer to refer to such a system as HSSI, as CDR block is just a part of the receiver.

Y. Fan, Z. Zilic, *Accelerating Test, Validation and Debug of High Speed Serial Interfaces*,
DOI 10.1007/978-90-481-9398-1_2,

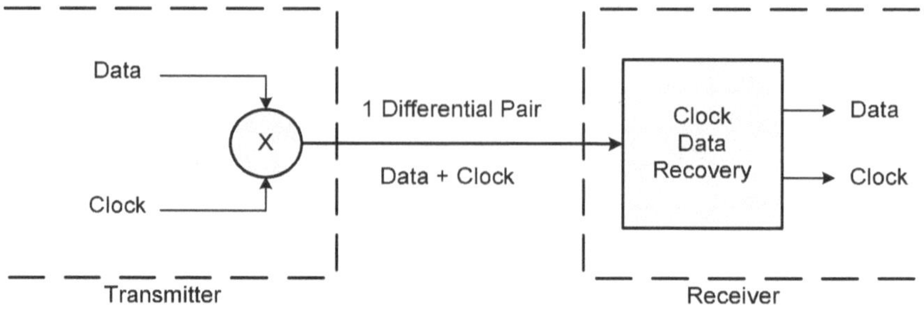

Fig. 2-1. CDR transmission mechanism

The CDR circuitry is nevertheless the key structure of an HSSI. At present, an HSSI that employs the CDR technology and differential signaling can provide communication at data rates above 10 Gbps. The roadmaps of many companies point to even higher data rates on each pair of wires. In addition, a wider pipe, or datapath, can be built by gluing multiple transceivers. Figure 2-2 shows such an example where 64 bits of data at 250 MHz are transmitted through 8 serial transceivers. Each transceiver works at 2.5Gbps, but as 8B10 encode/decode is commonly used to guarantee data transitions for clock recovery, each serial channel transmits actual data at 2 Gbps. Hence, the aggregated data rate is 16 Gbps.

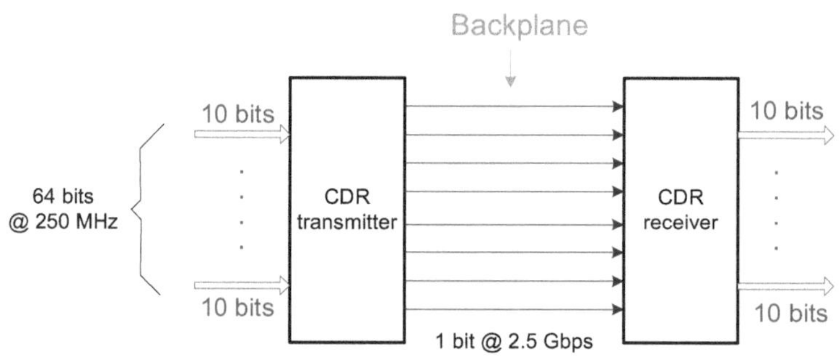

Fig. 2-2. Applications of multiple HSSIs

Besides the high-speed capability, the serial communication also simplifies routing among Integrated Circuits (ICs), or even parts of an IC. In parallel communication, data is usually transmitted one bit at a time down one wire. Routing for parallel communication is always challenging. For example, in an 8-bit parallel communication system, 8 wires or 16 wires (in the case of differential IOs) are needed for data signals and another one or two wires are needed for the clock signal. Routing 9 or 18 wires across a board, and keeping them all synchronized is very difficult and costly, especially for long distance connections.

In multi-gigabit HSSIs, differential I/Os are used so that two wires are needed for each connection, but the wire count is still much reduced relative to the parallel approach. Furthermore, in serial communication, the clock signal is embedded in the data and no clock skew exists. All these factors greatly simplify routing of serial communication. Because much fewer wires are used in a serial communication system compared to a traditional parallel one, it is further possible to put more circuitry on one die or in one package. Serial communication greatly relieves the package pin count bottleneck problem for SoCs.

Finally, a significant advantage of serial communication is in its lower energy consumption. Low energy consumption is facilitated by low-voltage differential signaling technologies, such as LVDS. The LVDS technology uses a constant-current line driver rather than a voltage-mode driver, so the supply current remains constant as the operating frequency increases, whereas the supply current for CMOS technology increases as the frequency increases. For a typical LVDS driver, as shown in Figure 2-3, the constant current is typically 3.5 mA [21]. The current passes to a resistor of about 100 Ohms (matched to the cable impedance) at the receiver end, generating a signal with amplitude around 350mV. The low power consumption of serial communication interfaces eliminates the need for either heat sinks or special packaging. Hence, serial communication reduces the system cost.

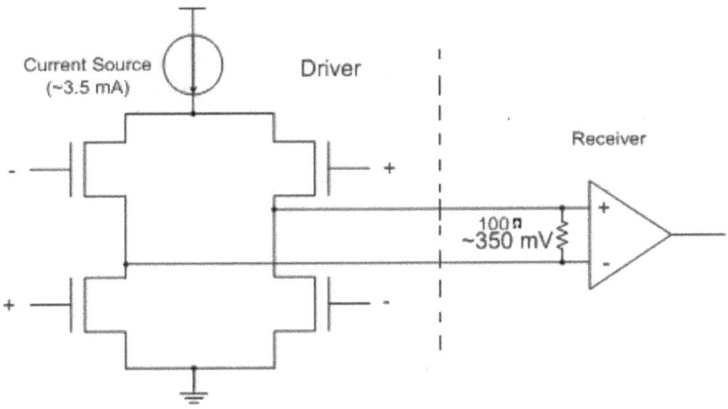

Fig. 2-3. Current-mode LVDS driver

Due to these advantages, multi-gigabit HSSIs are more and more widely used in high speed communications between devices, boards and systems. To address different applications, several HSSI standards have been developed, such as SATA [25], Fiber Channel (FC) [23] and 10 Gigabit Attachment Unit Interface (XAUI) [24]. Each standard specifies the detailed requirements of its functionalities and signal/device parameters, such as the data rate, amplitude level, slew rate and jitter. As an example, Table 2-1 lists the physical layer general specifications for SATA. Each device or system designed for the standard has to comply with these specifications.

Table 2-1. SATA Physical Layer General Specifications [22]

Parameter	Units	Limit	Gen1	Gen2	Gen3
Channel Speed	Gbps	Nom	1.5	3.0	6.0
FER*		Max	8.2e-8	8.2e-8	8.2e-8
T_{UI}, Unit Interval	ps	Min	666.4333	333.2167	166.6083
		Nom	666.6667	333.3333	166.6667
		Max	670.2333	335.1167	167.5583
F_{tol}, Tx freq. accuracy	ppm	Min	-350	-350	-350
		Max	350	350	350

* The Frame Error Rate is defined at the 95% confidence level

SATA is one of the most popular HSSI standards. It is the primary storage interconnect for PCs to connect the host system to peripherals, such as hard disk drives, solid state drives, optical drives and removable magnetic media devices [25]. SATA is an evolutionary replacement for the Parallel ATA interface. It can drastically increase the communication bandwidth and reduce the design cost compared to Parallel ATA. Even compared to other HSSI standards used in storage systems, SATA costs significantly less than SCSI or FC hard drivers. SATA market share has increased tremendously during the past few years: from 43% in 2006 to 97.7% in 2008 in the mobile PC market, and from 58.1% in 2006 to 99% in 2008 in the desktop PC market [26]. More than 1.1 billion SATA hard drivers have been shipped from 2001 to 2008 [27].

Therefore, it is especially beneficial to develop cost-efficient test methodologies for SATA. The experiments that will be described in this book concentrate on SATA devices, but the developed methodologies are equally applicable to other HSSI standards.

2.1.1 HSSI Structure

An HSSI consists of two parts: a transmitter (Tx) and a receiver (Rx), connected by transmission medium, such as cables or PCB traces. Figure 2-4 shows the block diagram of an HSSI. The transmitter and the receiver can function independently for half-duplex operation. They can also be combined for full-duplex operation.

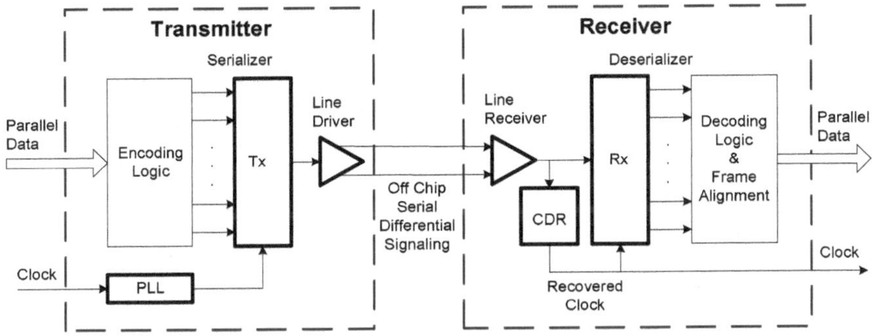

Fig. 2-4. Block diagram of an HSSI

The transmitter takes parallel data and converts it into a serial format. The PLL in the transmitter generates an internal high-speed clock for the serializer to provide the synchronization mechanism for the outgoing serial data. The differential line driver drives the serialized data into the transmission media. The receiver, on the other hand, accepts the high-speed serial data from the transmitter. The CDR in the receiver recovers a clock from the received serial data, and re-times the data using the recovered clock. Then the deserializer restores the re-timed serial data to the parallel format.

In the above transmission mechanism, the clock is embedded in the data signal, so only one differential data signal needs to be transmitted. The clock is then recovered from the serial stream by the CDR at the receiver. The CDR circuit extracts the clock information by monitoring the edge transitions of the received data. To ensure that the CDR circuit can function correctly, special encoding logic is needed in the transmitter to make sure that the transmitted data has enough transitions all the time.

One solution to guarantee the transitions is to encode the original data using a specialized encoder – most commonly, we will find an 8B10B encoder, because it encodes on a byte-by-byte basis. An 8B10B encoder converts 8-bit words into 10-bit words, so it can ensure that there are sufficient bit transitions, regardless of what pattern is transmitted. For example, in the 8B10B encoding scheme, there are four different symbols for the zero character; all the four symbols have transitions, but get interpreted as zero. In the receiver, the frame alignment block recognizes the word boundary and correctly restores the transmitted parallel sequences. Then the 8B10B decoding logic converts the 10-bit format to the original 8-bit format. The converted sequences are presented at the output ports of the receiver.

In the HSSI structure shown in Figure 2-4, the *Encoding Logic* block and *Decoding logic & Frame Alignment* block can be built with digital circuits; all other blocks can only be built with analog circuits. For HSSIs embedded in FPGAs, the analog blocks are usually hard cores; users can instantiate them and set some parameters, such as PLL frequencies and differential signaling formats. For the digital blocks, the users have the option to use Intellectual Property (IP) cores or develop their own designs. We will demonstrate how these blocks can be instantiated or built in Chapter 5.2.3.

In any communication system, including an HSSI system, we need physical medium to transmit the signal from the transmitter to the receiver. The medium may be PCB traces, cables, or optical fiber in serial communication. One essential feature of any such communication medium is that the transmitted signal can be additionally corrupted by a variety of possible mechanisms, such as medium loss and additive thermal noise.

2.1.2 BER Mechanisms

In the serial communication system, the transmitter, the receiver or the transmission media can introduce distortion or cause bit errors. The correctness and performance of communication interfaces depend on many design choices, such as the CDR mechanism, the PLL bandwidth, the method of encoding/decoding and the transmitter power. As a measure of how well the overall communication system performs, BER is the probability of a bit error at the output of the receiver, compared to the input of the transmitter [28].

By definition, BER is derived by calculating the ratio of the number of erroneous bits to the number of transmitted bits. The concept is very simple, but there are a few issues we need to consider when performing the BER measurement, such as how many bits need to be transmitted and how many bit errors need to be captured in order to get a reliable BER test result. If we transmit 10^{12} bits and get one erroneous bit, can we claim that the BER performance of the system is 10^{-12}? If we transmit another 10^{12} bits, will we also get one and just one erroneous bit? One would most likely not be able to claim so. The BER *confidence level* is used to resolve this dilemma and to define how much confidence in the test result there is.

For a given digital communication system or component, there usually is a minimum specification for the BER - $p(e)$. In practice, $p(e)$ is often estimated by calculating the ratio of detected bit errors (*l*) to total bits transmitted (*n*) in a fixed length test sequence. If we denote the ratio by $p'(e)$, $p'(e)$ is only an estimation of $p(e)$. The estimation accuracy improves with the increase of the number of transmitted bits. As shown in the following equation, $p'(e)$ only equals to $p(e)$ if an infinite number of bits are transmitted.

$$p'(e) = \frac{l}{n} \xrightarrow{n \to \infty} p(e)$$

However, it is impossible to transmit an infinite number of bits to get $p(e)$ in real BER testing because the test time would be infinite. The actual number of transmitted bits depends on the desired BER confidence level. The BER confidence level will require the use of a probability that the actual $p(e)$ is better than a specified BER level y (such as 10^{-12}).

The Confidence Level (CL) is formally expressed as a conditional probability for a given BER level upon discovering the corresponding number of erroneous bit transitions

$$CL = p[p(e) < y \mid l, n]$$

where y is a specified BER level, and $\mid l, n$ denotes a condition that n bits are transmitted and l bits of errors are detected.

The confidence level is based on a set of BER measurements. One interpretation of the confidence level is that, if the BER test is repeated many times and the value $p'(e) = l/n$ is recomputed each time, we expect $p'(e)$ to be better than the BER level y for CL (in percent) of the measurements. To measure BER with a constant confidence level, we need to use a variable-length test sequence [29], [30]. The BER confidence level can be calculated based on the binomial distribution function [31], [32]. The binomial distribution function models events that have only two possible outcomes, and is generally written as

$$p_n(k) = \binom{n}{k} p^k q^{n-k}, \text{ where } \binom{n}{k} = \frac{n!}{k!(n-k)!}$$

where $p_n(k)$ is the probability that k events (i.e., bit errors) occurs in n trials (i.e., bits transmitted), p represents the probability that an event occurs in a single trial (i.e. a bit error), and q represents the probability that the event does not occur in a single trial. According to these denotations, we have $p + q = 1$.

Figure 2-5 plots the binomial distribution with $n = 10^8$ and $p = 10^{-7}$. If we treat the n as the total number of bits transmitted, p as the BER, and $p_n(k)$ as the probability that k bit errors will occur, we can use the distribution to calculate the BER confidence level. We are interested in the probability that N or fewer bit errors occur in n transmitted bits. The probability is the Cumulative Distribution Function (CDF) of the binomial distribution and is expressed as

$$p(e \le N) = \sum_{k=0}^{N} p_n(k) = \sum_{k=0}^{N} \frac{n!}{k!(n-k)!)} p^k q^{n-k}$$

Then, the confidence level can be expressed as:

$$CL = 1 - \sum_{k=0}^{N} [\frac{n!}{k!(n-k)!}] p^k (1-p)^{n-k}$$

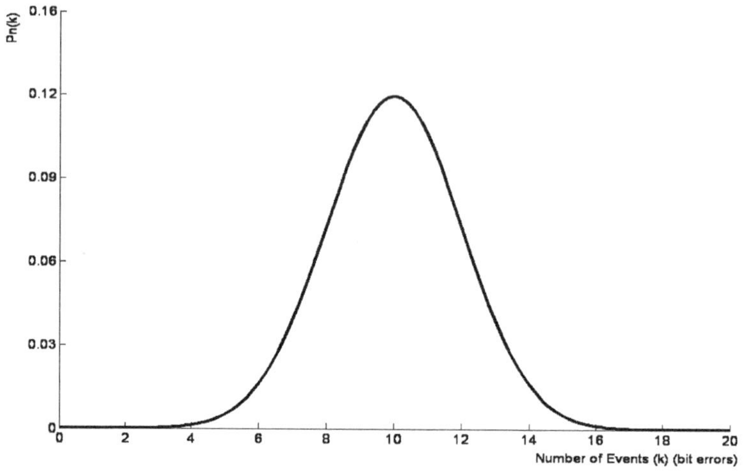

Fig. 2-5. Graph of the binomial distribution ($n = 10^8, p = 10^{-7}$)

To qualify a BER p, we need to determine how many bits n must be transmitted with N or few errors. We can first choose a hypothetical value of p and a desired CL, then solve the above CL equation to determine n and N to prove the hypothesis. It is difficult to directly solve n and N. One solution is to use Poisson theorem [31] to simplify solving n and N. Poisson theorem provides a conservative estimate of the binomial distribution function, and is expressed as

$$p_n(k) = (\frac{n!}{k!(n-k)!})p^k q^{n-k} \xrightarrow{n\to\infty} \frac{(np)^k}{k!} e^{-np}$$

An example of the solutions for N and n is listed in Table 2-2 [33]. In this system, p is specified to 10^{-10} and the confidence level CL is set to 99%. For various bit errors of N, the corresponding required transmitted bits of n are solved. As can be seen from the table, we will have a 99% confidence level in claiming that $p(e) < 10^{-10}$ in a 2.5Gbps system if no erroneous bit is detected in 18.5s of testing, one erroneous bit occurs in 26.7s, or two erroneous bits occur in 33.7s. Hence, given a desired confidence level, we can select the sufficient duration of the bit stream from such a table and depending on how many erroneous bits were detected.

Table 2-2. An Example of BER Estimation (CL=99% and $p = 10^{-10}$)

Bit Errors N	0	1	2	3	4
Required bits n	$4.61*10^{10}$	$6.64*10^{10}$	$8.40*10^{10}$	$1.00*10^{11}$	$1.16*10^{11}$
Test time @ 2.5Gbps (s)	18.5	26.7	33.7	40.2	46.6

The test time t can be calculated using Gaussian error distribution:

$$t = \frac{\ln(1-CL)}{p*r}$$

where CL is the confidence level, p is the upper bound of BER and r is the data rate. Figure 2-6 shows the relationship between the test time and the confidence level. If we want to achieve a higher confidence level, the test time must increase. We cannot achieve 100% confidence level in BER testing because that would require infinite test time. The BER measured by BERT equipment is only an estimate of the true BER. In practice, we are hence forced to make tradeoffs between the confidence level and the test time.

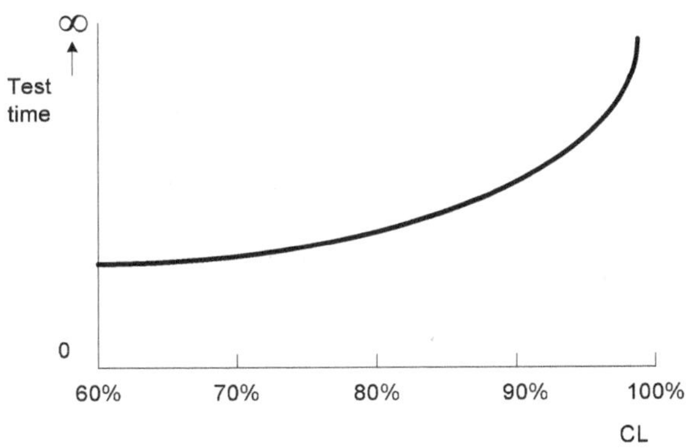

Fig. 2-6. Test time vs. BER confidence level

Table 2-2 also shows that we need to give some margin for a measured BER if we use only a few bit errors to qualify BER performance. For example, if we measure 3 bit errors out of 10^{11} transmitted bits, we should not think that the BER performance is $3.33*10^{-11}$. It is only reasonable to say that we have 99% confidence level that the BER performance is better than 10^{-10}. As a rule of thumb, we normally need at least $10*(1/p)$ transmitted bits in order to qualify p BER performance. As shown in Table 2-2, to qualify 10^{-10} BER performance, we can tolerate up to 3 bit errors if we transmit 10^{11} bits. This provides us guidelines on how to efficiently and confidently test the BER for jitter tolerance qualifications discussed in Chapter 3.

2.1.3 Jitter and Noise Impacts to BER

In serial communication systems, there are various signal formats in time domain. The most commonly used format is Non Return Zero (NRZ). Non Return Zero Inverted (NRZI) and Return Zero (RZ) are also used in some systems. In NRZ-encoded format, the binary information digit 1 is encoded as a high signal represented by "1", and the binary information digit 0 is encoded as a low signal represented by "0". The ideal NRZ signal can be represented by a trapezoidal waveform as shown in Figure 2-7. The waveform consists of four components: high level "1", low level "0", rising edge (0 to 1 transition) and falling edge (1 to 0 transition).

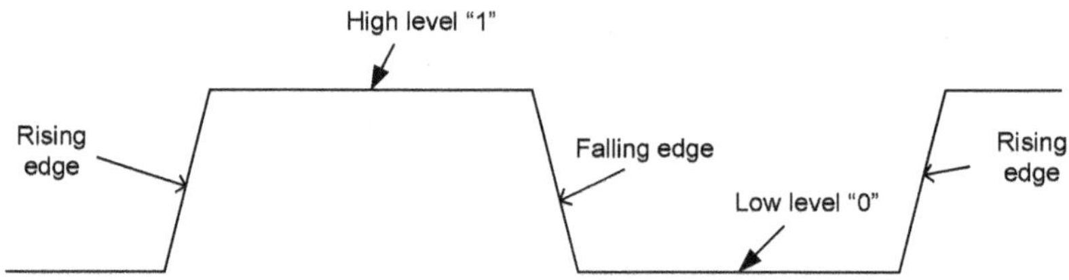

Fig. 2-7. Ideal digital signal

When the ideal signal is transmitted, it gets contaminated by a variety of physical processes that we collectively refer to as *noise*. The deviation of a noise-contaminated signal from its ideal position can be viewed from two aspects: time-deviation (jitter) and amplitude-deviation (amplitude noise). Figure 2-8 illustrates the two deviations: Δt and Δv.

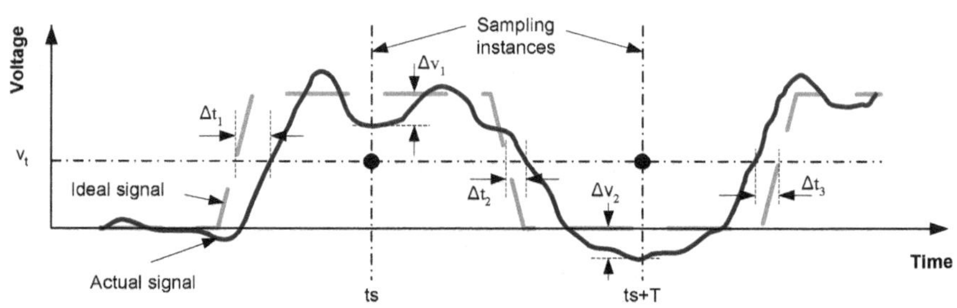

Fig. 2-8. Timing and amplitude deviations in an actual data signal.

At the receiver side, the CDR samples the actual data signal at sampling instance t_s and compares the sampled value with a threshold voltage V_t. If the value is bigger than V_t, logic "1" is received; otherwise, logic "0" is received. An ideal receiver samples data in the middle of each data bit. Without amplitude noise, the receiver can always correctly recover the transmitted bit. Under the presence of jitter and noise, the transition edge of the signal can fluctuate horizontally across the sampling point (along the time axis), and the signal voltage can fluctuate vertically at the sampling point (along the voltage axis). Both the time deviation and amplitude deviation can cause a bit error – bit "0" is received as bit "1" or bit "1" is received as bit "0".

If we ignore the setup time and hold time requirements [109], timing jitter can cause bit errors in the following two conditions:

- For logic "1", the rising edge lags behind the sampling instance or the falling edge is ahead of the sampling instance
- For logic "0", the falling edge lags behind the sampling instance or the rising edge is ahead of the sampling instance

Amplitude noise causes bit errors in the following two situations:

- For logic "1", the voltage level at the sampling instance is smaller than the threshold voltage
- For logic "0", the voltage level at the sampling instance is bigger than the threshold voltage

The timing and amplitude noise impacts on BER can be characterized by two parameters – Jitter and Signal-to-Noise Ratio (SNR). The SNR quantifies the amplitude noise while jitter represents the timing deviation.

2.2 Timing Jitter

2.2.1 Jitter Overview

Jitter is the deviation of a signal from its ideal timing. There are several different types of jitter, which all get superimposed into the Total Jitter (TJ). The simplest view is that the TJ is composed of Random Jitter (RJ) and Deterministic Jitter (DJ) [34], [35], [36]. RJ is caused by random events and is usually characterized statistically by a Gaussian distribution, which can be quantified by a mean and a standard deviation. Because a Gaussian distribution is unbounded (its peak-to-peak value approaching infinity), the peak-to-peak value is dependent on the total sample size. Therefore, the peak-to-peak value of RJ is related to the BER and TJ is defined at a certain BER level.

DJ is caused by deterministic events and has a bounded peak-to-peak value. DJ can be decomposed into Periodic Jitter (PJ), Bounded Uncorrelated Jitter (BUJ) and Data Dependent Jitter (DDJ) [37]. PJ is caused by repetitive noise sources, such as clock signals and oscillators. BUJ is usually caused by coupling, such as from adjacent signal paths and on-chip logic switching. DDJ depends on the bit pattern in the signal under test and can be further classified into Duty Cycle Distortion (DCD) and Inter-Symbol Interference (ISI). DCD is caused by an imbalance in the drive circuit, which generates a signal having unequal pulse widths for high and low logic values. ISI is caused by frequency related losses in the signal path, such as those caused by the bandwidth limitation. Figure 2-9 shows the relationship of all the jitter components.

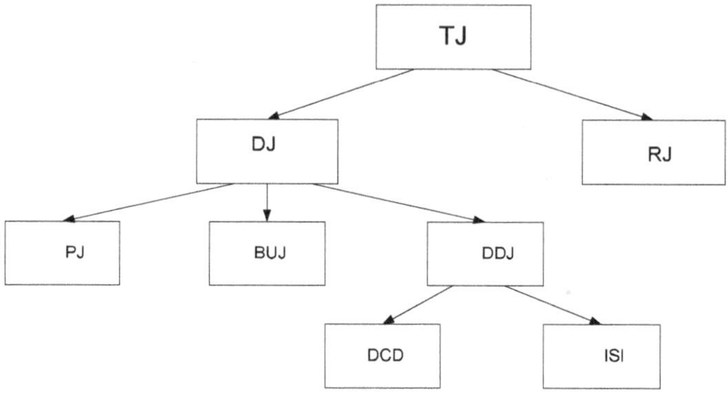

Fig. 2-9. Jitter components

It is important to appreciate that the different jitter components exhibit widely different behavior, as given, for instance, by a Probability Density Function (PDF) characterizing the probability distribution of a jitter. Figure 2-10 illustrates the typical PDF functions for PJ, RJ, DCD and ISI. Only RJ is unbounded, most often associated with Gaussian distribution, while DCD and ISI are dependent on the data pattern and are mostly thought of as having a discrete distribution. Finally, as PJ is caused by a combination of repetitive events, the distribution is continuous and largely independent of the data stream.

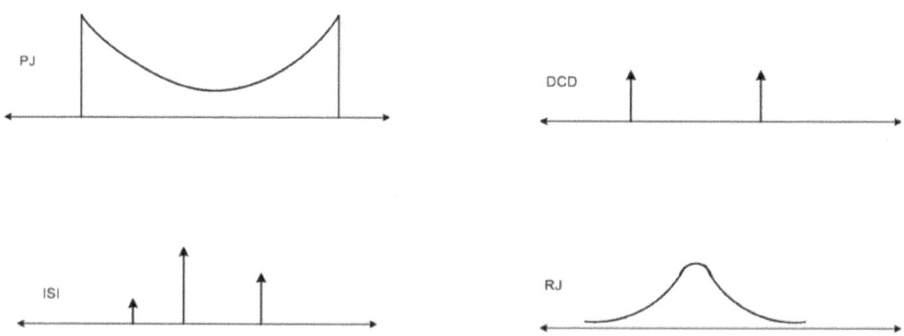

Fig. 2-10. Jitter components

Most communication standards, such as SATA, Fiber Channel and XAUI, specify jitter in terms of DJ and TJ as separate specifications [22], [23], [24]. Table 2-3 summarizes the HSSI transmitter jitter specifications for the SATA [22] at 10^{-12} BER. The normal data rate is 1.5Gbps for Gen1, 3.0Gbps for Gen2 and 6.0Gbps for Gen3. Table 2-4 summarizes the SATA specifications of the lab-sourced signals for HSSI receiver jitter tolerance testing. As we can see, the maximum jitter the receiver should tolerate is higher than the maximum jitter allowed from the transmitter output. For example, the receiver TJ tolerance specification for Gen2 is 0.60UI while the transmitter jitter specification is 0.37UI. This is understandable because the transmission medium between the transmitter and the receiver may introduce extra jitter.

Table 2-3. Transmitter Jitter Specifications for SATA

Parameters	Units	Limit	Gen1	Gen2	Gen3
TJ, f_{BAUD}/500	UI	Max	0.37	0.37	--
DJ, f_{BAUD}/500	UI	Max	0.19	0.19	--
TJ*	UI	Max	--	--	RJp-p + 0.34
RJ*	UI	Max	--	--	0.18p-p

* More detailed test conditions are defined in [22]

Table 2-4. Receiver Jitter Tolerance Specifications for SATA

Parameters	Units	Limit	Gen1	Gen2	Gen3
TJ, f_{BAUD}/500	UI	Max	0.60	0.60	--
DJ, f_{BAUD}/500	UI	Max	0.42	0.42	--
TJ after CIC, JTF defined	UI	Max	--	--	0.60
RJ before CIC, MFTP JTF defined	UI	Max	--	--	0.18p-p (2.14ps, 1 sigma)

2.2.2 *Jitter and BER*

Jitter can cause bit errors in serial communication when the recovered clock retimes the serial data signal. Figure 2-11 illustrates the relationship between jitter and BER. Ideally, the data is always sampled in the mid-bit, sampling instance t_s = UI/2 in Figure 2-11(a), where UI is the Unit Interval (period) of the signal. This is usually true if the jitter frequency is within the bandwidth of the CDR because the sampling clock is recovered from the data signal and the clock can track the in-band jitter. However, for out-of-band jitter, the sampling clock cannot track the data any more and the jitter can cause bit errors.

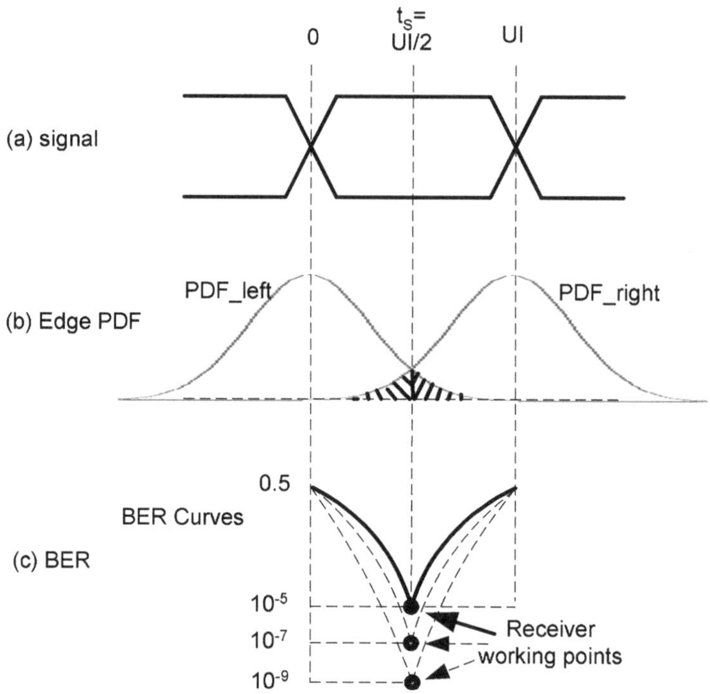

Fig. 2-11. Jitter and BER in the receiver

Figure 2-11(b) shows an example of the jitter profile. Here, we assume that the signal edge transition is disturbed by RJ, and RJ alone. The RJ is further assumed to be Gaussian. The Probability Density Function (PDF) of a zero-mean Gaussian variable is

$$p(x) = \frac{1}{\delta\sqrt{2\pi}} e^{-x^2/2\delta^2}$$

where δ is the standard deviation.

One important function used to characterize the Gaussian distribution is the *Q factor*, which represents the area under the tail of the Gaussian PDF. The Q factor is widely used for computing the error probability in communication systems [38]. Normalized to zero mean and unit variance, $Q(x)$ is defined as

$$Q(x) = \frac{1}{\sqrt{2\pi}} \int_x^{\infty} e^{-t^2/2} dt, \quad x \geq 0$$

$$= \frac{1}{2} erfc(\frac{x}{\sqrt{2}}) \qquad (2\text{-}1)$$

where *erfc*(x) denotes the complementary error function, defined as

$$erfc(x) = \frac{2}{\sqrt{\pi}} \int_x^{\infty} e^{-t^2} dt \qquad (2\text{-}2)$$

As shown in Figure 2-11(b), the PDF of the left and right edge transitions of the data bit can be expressed by

$$p_{\_left}(x) = \frac{1}{\delta\sqrt{2\pi}} e^{-x^2/2\delta^2}$$

$$p_{\_right}(x) = \frac{1}{\delta\sqrt{2\pi}} e^{-(x-UI)^2/2\delta^2}$$

Bit errors may occur when the left edge occurs after the t_s (illustrated by back slash in Figure 2-11(b)) or the right edge occurs before the *ts* (illustrated by forward slash in Figure 2-11(b)). Assuming there is the uniform bit distribution, i.e., that there is a 50% chance of errors in these two cases, the BER can be obtained by calculating the probability of the two erroneous cases:

$$BER(t_s) = 0.5 * (\int_{t_s}^{\infty} p_{\_left}(t)dt + \int_{-\infty}^{t_s} p_{\_right}(t)dt$$

$$= 0.5 * (\int_{t_s}^{\infty} \frac{1}{\delta\sqrt{2\pi}} e^{-\frac{t^2}{2\delta^2}} dt + \int_{-\infty}^{t_s} \frac{1}{\delta\sqrt{2\pi}} e^{-\frac{(t-UI)^2}{2\delta^2}} dt)$$

$$= \int_{t_s}^{\infty} \frac{1}{\delta\sqrt{2\pi}} e^{-\frac{t^2}{2\delta^2}} dt \qquad (2\text{-}3)$$

According to Equation (2-2), Equation (2-3) can be rewritten as

$$BER(ts) = 0.5 * erfc(\frac{t_s}{\delta\sqrt{2}}) \qquad (2\text{-}4)$$

By substituting δ with the Root-Mean-Square (RMS) value of the RJ, $RJ_{RMS.}$, Equation (2-4) becomes

$$BER(ts) = 0.5 * erfc(\frac{t_s}{RJ_{RMS}\sqrt{2}}) \qquad (2\text{-}5)$$

Equation (2-5) directly links the jitter to BER. The BER and jitter relationship can further be transferred to the BER and Q factor relationship. As shown in Figure 2-11(c), the receiver works at the crossing points of the bathtub curves. Therefore, we have

$$t_s = UI/2 \tag{2-6}$$

$$UI = DJ + 2Q * RJ_{RMS} \tag{2-7}$$

where Q is $Q(x)$ defined in Equation (2-1) with $x = BER$ [34].

By substituting t_s and RJ_{RMS} in Equation (2-5) according to s (2-6) and (2-7), we have

$$BER = 0.5 * erfc(\frac{Q}{\sqrt{2}}) \tag{2-8}$$

or

$$Q = \sqrt{2} * erfc^{-1}(2 * BER) \tag{2-9}$$

Equations (2-8) and (2-9) directly link BER and Q factor. If we know one parameter, the other can be calculated accordingly. We would need these two equations for our jitter tolerance extrapolation.

In the above analysis, we ignore the DJ effect. This is reasonable since the bit errors are caused by RJ at low BER regions. Because the DJ is bounded, it only adds offsets to a bathtub curve; it does not change the shape of the lower part of the bathtub curve [34].

2.2.3 Jitter Testing

As we can see, jitter can cause bit errors. From the receiver point of view, there are two possible factors that can cause excessive bit errors. One is that there is excessive jitter in the received signal, which means that the transmitter generates too much jitter (assuming there are no issues with the transmission media). Another factor is that the minimum jitter that the receiver can tolerate at a certain BER level is lower than what is expected. To quantify the two factors, the *transmitter jitter* specification defines the maximum jitter the transmitter is allowed to generate, while the *receiver jitter tolerance* defines the minimum jitter the receiver should tolerate to achieve a certain BER level. Therefore, in HSSI jitter compliance testing, we qualify the transmitter jitter and the receiver jitter tolerance performance.

Because the jitter is usually defined at 10^{-12} or lower bit error rates, direct measurements are too time consuming to conduct in most cases. Considering the BER performance of an HSSI is related to jitter according to Equation (2-5) or related to Q factor according to Equation (2-8), Q factor can be used to accelerate the BER and jitter measurement. One important assumption for the Q factor approach is that RJ is Gaussian. A Gaussian distribution is theoretically unbounded and can be characterized by its standard deviation. For example, a Gaussian distribution on average will exceed a span of 14 times its standard deviation one time for every 10^{12} samples, which can interpreted as Q factor is 7 (14/2 =7) at 10^{-12} BER. If we know the DJ and the RJ, we can estimate the transmitter TJ as follows:

$$TJ = DJ + 2Q * RJ$$

where the Q factor is 7 at 10^{-12} BER.

The above Q factor based transmitter TJ estimation is much faster than the direct measurements at lower BERs. We can also use the Q factor to accelerate the receiver jitter tolerance testing, as will be discussed in Chapter 3. Even though the Q factor is widely used in the industry [34], [140], the measurement accuracy may be affected if the RJ distribution deviates from the true Gaussian distribution [141]. When the Power Spectrum Density (PSD) of the RJ is not white, its PDF can not be described as Gaussian, such as the high-order jitter discussed in [7]. For non-Gaussian distributions, we cannot use the Equation (2-3) to calculate the BER as shown in the overlap area in Figure 2-11(b).

However, the Q factor based approaches in this book are still valid because RJ is primarily due to the thermal noise in electronic components, which has a white power spectrum density and whose PDF is Gaussian. Further, the fundamental fact is that the random noise is caused by a combination of various uncorrelated random noise sources, and by central limit theorem, the distribution of the aggregated noise approaches Gaussian distribution. Even though there are some non-Gaussian RJ sources, such as 1/f noise and shot noise [7], our eye mask testing approach discussed in Chapter 4.3.5 will provide a mechanism to deal with this case.

For HSSI receiver jitter tolerance testing, we need to stress the receiver with a pre-determined amount of jitter according to the related specification and then qualify whether the BER performance of the receiver is better than the specified BER level. Because the BER is usually defined at 10^{-12}, it is too time consuming to conduct direct measurements in most cases. To accelerate the jitter tolerance qualification, we stress the receiver with controllable amounts of injected jitter. By varying the injected jitter in the input signal, we can make the receiver work at different higher BER levels, which are much less time-consuming to test. The jitter tolerance performance at low BER levels can then be extrapolated according to the measured jitter vs. BER relationship at high BER levels. The details of the jitter tolerance testing will be discussed in Chapter 3.

2.3 Amplitude Noise

2.3.1 BER and SNR

As discussed in section 2.1.3, amplitude noise can cause bit errors when the noise exceeds a certain level. The BER characterizes the probability of bit errors and the SNR defines the ratio of the signal amplitude over the noise amplitude. The relationship between BER and SNR can be derived from the Additive White Gaussian Noise (AWGN) communication channel model [38]. The basic components of a communication system include a transmitter, a receiver and a communication channel. One problem associated with the channel is that it corrupts the transmitted signal in a random manner. The AWGN model is predominantly used to analyze this problem because the noise is predominantly white: its PSD is flat and its distribution is Gaussian [7], [34], [140]. The mathematical model of the AWGN channel is shown in Figure 2-12 [38].

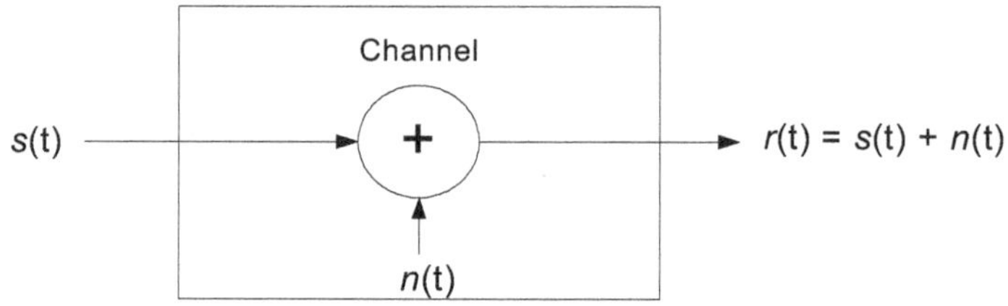

Fig. 2-12. AWGN channel model

In the AWGN model, the transmitted signal $s(t)$ is corrupted by noise $n(t)$. The model can be expressed by

$$r(t) = s(t) + n(t).$$

The noise is introduced by the channel, as well as by electronic components, including amplifiers at the receiver. This type of noise is most often characterized as thermal noise, or statistically as Gaussian noise. Its PDF is expressed by

$$p(x) = \frac{1}{\delta\sqrt{2\pi}} e^{-(x-m_x)^2/2\delta^2}$$

where m_x is the mean and σ^2 is the variance of the Gaussian random variable.

In general, BER is a function of the characteristics of the channel (i.e., amount of noise), the type of waveforms used to transmit information over the channel, the transmitter power, the timing jitter, and the method of demodulation and decoding. In baseband transmission, the data and clock are combined on the transmitter side and transmitted as digital waveforms.

One commonly used baseband scheme is NRZ CDR encoding, which has been discussed in Section 2.1. The receiver in the system samples the received bit stream at an appropriate time point and decides whether the sampled value represents a binary one or zero. Careful timing extraction leads to a reduction in the number of transmission errors. A matched filter is a linear system that significantly alters the shape of both the signal and the noise in a way that increases the SNR. Figure 2-13 shows the structure of a binary matched filter receiver [38], [45]. We next introduce the "one or zero" decision principle and set the stage to evaluate the BER performance of the binary matched filter receiver.

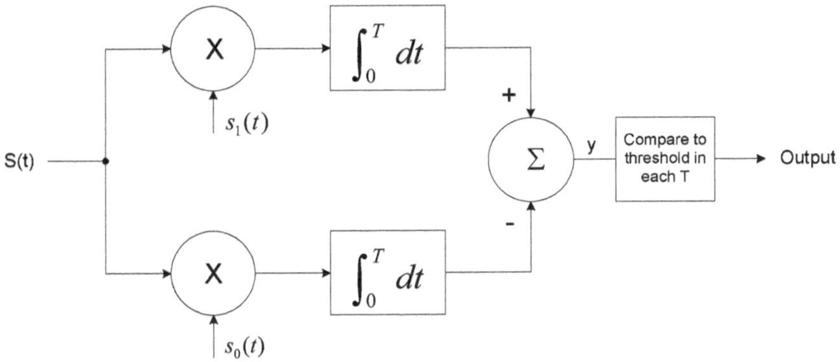

Fig. 2-13. Binary matched filter receiver

The receiver consists of two filters, one matching to $S_0(t)$ and the other matching to $S_1(t)$. Each filter consists of a multiplier and an integrator. The receiver compares the output of the two filters. The output of each integrator is a number composed of a deterministic part (due to the signal) and a random part (due to the additive noise). The additive noise is assumed to be zero-mean Gaussian and its frequency spectrum is flat. Suppose the input signal to the receiver is $s_i(t) + n(t)$, where i is either zero (binary 0 is transmitted) or unity (binary 1 is transmitted). The comparator input is

$$y = \int_0^{T_s} s_i(t)[s_1(t) - s_0(t)]dt + \int_0^{T_s} n(t)[s_1(t) - s_0(t)]dt$$

The value of y consists of two integrals. The average of the second integral is zero since the noise is zero mean. Therefore, due to the non-biased noise, the mean value of y is given by

$$m_y = \int_0^{T_b} s_i(t)[s_1(t) - s_0(t)]dt$$

and the variance of y is given by

$$\sigma_y^2 = E\{[y - m_y]^2\}$$

$$= E\{[\int_0^{T_b} n(t)[s_1(t) - s_0(t)]dt]^2\}$$

$$= E\{\int_0^{T_b}\int_0^{T_b} n(t)n(v)[s_1(t) - s_0(t)][s_1(v) - s_0(v)]dtdv\}$$

Because the noise is assumed to be white, with power spectral density

$$G_n(f) = N_o/2,$$

the autocorrelation of the noise is the inverse Fourier transform of the power spectrum, or

$$R_n(t) = N_o\,\delta(t)/2.$$

According to this identity and the relationship between the expectation and autocorrelation

$$E\{n(t)n(v)\} = R_n(t - v),$$

we have

$$\sigma_y^2 = \{\int_0^{T_b}\int_0^{T_b} \frac{N_o}{2}\delta(t - v)[s_1(t) - s_0(t)][s_1(v) - s_0(v)]dtdv\}$$

$$= \frac{N_0}{2}\int_0^{T_b} [s_1(t) - s_0(t)]^2\, dt$$

According to the mean and variance of y, we can get the probability density of y. As shown in Figure 2-14, the probability density fits into one of the two PDFs labeled with $p_0(y)$ -- 0 being transmitted, and $p_1(y)$ -- 1 being transmitted. The two functions have the same variances but different mean values, m_0 and m_1 respectively.

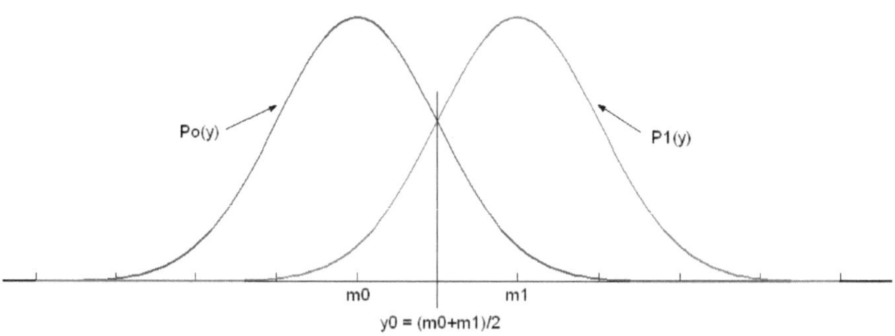

Fig. 2-14. Probability densities of y

The comparator threshold is chosen as the two-PDF cross point labeled as y_0. If y is greater than y_0, we assume that $s_1(t)$ is being sent; otherwise, $s_0(t)$ is being sent. Because of the symmetry, the threshold y_0 is the midpoint between the mean values of the two PDFs:

$$y_0 = \frac{m_1 + m_0}{2}$$

$$= \frac{1}{2}\int_0^{T^b} [s_1(t) + s_0(t)][s_1(t) - s_0(t)]dt$$

$$= \frac{1}{2}\int_0^{T_b} [s_1{}^2(t) - s_0{}^2(t)]dt$$

Because the integral of the square of the signal is the signal energy over the bit period, the threshold becomes

$$y_0 = \frac{E_1 - E_0}{2}$$

where E_1 and E_0 are denoted as the energy for the two signals $s_1(t)$ and $s_0(t)$, respectively.

According to the two PDFs shown in Figure 2-14, we can calculate the error probability of the receiver. The probability of mistaking a transmitted 1 for a 0 is the integral of $p_1(y)$ between [-∞, y_0], and the probability of mistaking a transmitted 0 for a 1 is the integral of $p_0(y)$ between [y_0, ∞]. Hence, the error probability is given by the areas under the tails of the PDFs. Assuming that the two signals, $s_0(t)$ and $s_1(t)$, have equal energy, the probability of an error is

$$P_e = \frac{1}{\sqrt{2\pi}\sigma}\int_0^{\infty} \exp[\frac{(y - m_0)^2}{2\sigma^2}]dy$$

$$= Q(\sqrt{\frac{E(1-\rho)}{N_o}}) \qquad (2\text{-}10)$$

where $Q(\ )$ is the Q factor (discussed in Chapter 2.2.2), E is the average energy of the two signals, ρ is the correlation coefficient of the two signals, and N_0 is the noise power per Hz. E, N_0 and ρ are defined as

$$E = \frac{E_0 + E_1}{2}$$

$$N_o = \frac{\sigma^2}{f_m} \quad (f_m \text{ is the bandwidth})$$

$$\rho = \frac{\int_0^{T_b} s_0(t) s_1(t) dt}{E}$$

As can be seen from Equation (2-10), the BER is determined by three factors: the average energy per bit E, the correlation of the two signals ρ, and the noise power per Hertz N_0.

To understand the relationship between BER and E/N_o, we consider three cases where the correlation coefficient ρ is different.

The first case assumes that $s_0(t) = s_1(t)$. Then the correlation coefficient ρ equals to 1. According to Equation (2-10), the BER is

$$P_e = Q(0) = \frac{1}{2}$$

This result can be easily explained. Since the same signal is used to transmit both 0 and 1, there is actually no information provided and the receiver can only randomly guess the received information.

The second case assumes that $s_0(t) = -s_1(t)$. In this case, the correlation coefficient ρ equals to -1, and the BER is

$$P_e = Q(\sqrt{\frac{2E}{N_o}})$$

The third case assumes that $s_0(t) = 0$ and $s_1(t) = 1$. Because only a single power supply is needed, most digital communication systems, including HSSIs, use this format. In this case, the correlation is $\rho = 0$, and the BER becomes

$$P_e = Q(\sqrt{\frac{E}{N_o}})$$

or

$$BER = Q(\sqrt{SNR}) \tag{2-11}$$

Figure 2-15 plots the relationship between the BER and the SNR for the above three values of correlation: -1, 0 and 1.

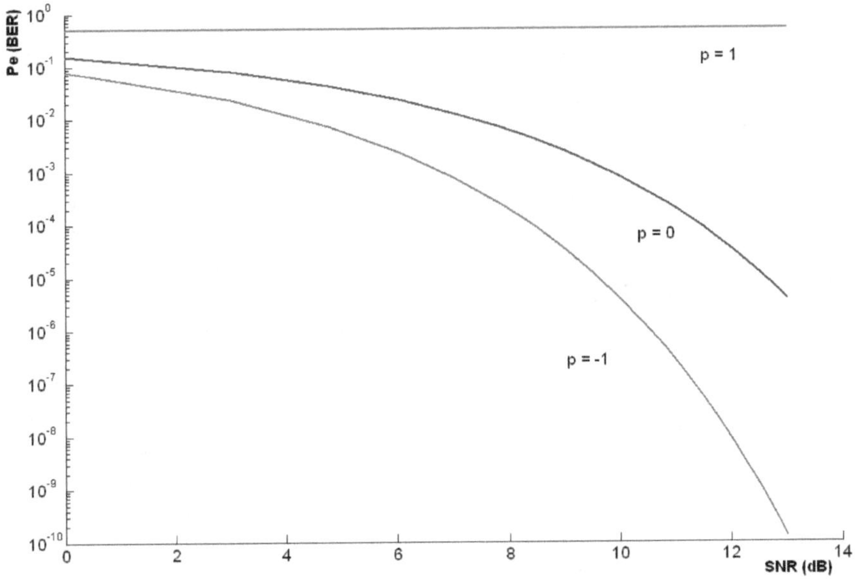

Fig. 2-15. BER vs. SNR for baseband transmission

If a Gaussian noise at the input of the receiver is the dominant cause of bit errors, according to Equation (2-11) we can get higher BER by reducing the SNR. In simulation or real testing, the SNR can be reduced to a known quantity by adding a controllable amount of noise to the test signal. Therefore, we need a high-quality noise generator in order to control the SNR to be able to perform high-quality testing procedures.

2.3.2 Simulation and Emulation

Usually, the BER performance of a communication system is first evaluated by software simulation in the early development stage. Simulation tools such as MATLAB and Simulink [46] are therefore used for pre-silicon evaluation. In software simulations, each component of a communication system, including the communication channel, is represented by a software model that exhibits the characteristics of the represented component. In software-based BER testing scheme, a communication channel is built using a software channel model. The BER performance of the communication system can be easily evaluated by running the simulations under different conditions of SNR.

Although software simulations are easy to set up, they are very time consuming for BER evaluation. The execution is typically done using workstation CPU processors or using acceleration methods. The execution speed depends on the level of abstraction of the simulation models. Due to the vast amounts of data and run-time overhead, simulations generally are only suitable for the evaluation of a system with low BER performance (such as BER $>10^{-6}$). For example, 10^{9} calculation iterations are needed to get an accurate (±3.3%) estimation of a BER around 10^{-6} [47]; a simulation of BER=10^{-8} with 10 errors takes days on a personal computer equipped with a 1 GHz Pentium 4 processor. In contrast, acceptable BER in commercial communication systems (e.g. data transmission) go below 10 errors out of 10^{9} transmitted bits in many cases. Moreover, many design variables, such as sampling frequency, digital format, etc., have to be optimized while satisfying the best trade-off between performances and complexity, which would further lengthen the simulation process.

In order to speed up the BER evaluation process, performing direct hardware simulation, i.e., emulation is proposed. As an alternative to simulation, emulation utilizes FPGAs to re-target all or part of a design. Many software tools and dedicated hardware [48], [49] have been developed with the aim of automating this re-targeting process. In emulation, performance evaluation takes place in hardware, rather than in the virtual environment of a simulator. A hardware-based solution is commonly 100,000 to 1 million times faster than the best simulation software at the same abstraction level [50].

Emulation also makes it possible to run a design at a real time system. This feature is especially important for applications such as the video/audio, where the final output needs to be observed in real time due to the subjective nature of the receiver (e.g. the human eye/ear). If such systems run in real time, the performance of the system can be evaluated on the fly. Otherwise, large test vectors need to be captured and replay mechanisms have to be created, which is much more time-consuming and the evaluation result is less reliable.

Overall, emulation can greatly reduce the evaluation time. Additionally, its real time test capacity can cover a much larger set of test conditions than simulation and hence emulation can enhance the evaluation quality. For these reasons, emulation has been widely used for performance evaluation [51], [52], [53], [54].

2.3.3 AWGN Emulation

As discussed in Chapter 2.3.1, the BER performance of a communication system is closely related to the SNR. There exist instruments for BER testing [55], [56], but few integrate an AWGN emulator. Therefore, such testers are difficult to set up for BER testing under the presence of noise. An AWGN generator is needed in order to perform BER performance evaluation under noise conditions.

AWGN generation usually utilizes a variety of statistical techniques. The implementations of AWGN generators are almost always based on transformations or operations performed on uniform random variables [57], [58], [59]. There are a few publications on generating AWGN in digital hardware. The most relevant publications in this area are [47], [60], [61], [62] and [63], which implement AWGN generators in FPGAs.

In this book, we propose a novel scheme to implement AWGN generators in hardware. We also explore the advantages and applications of our method in BER testing. More details are presented in Chapter 6.

3 Accelerating Receiver Jitter Tolerance Testing on ATE

Abstract *This chapter details the approaches that help to accelerate the receiver jitter tolerance testing thousandfold. According to the receiver characteristics and the test setup, we first propose a jitter tolerance extrapolation algorithm. Based on this algorithm, we then propose acceleration schemes for production test, as well as for characterization and silicon debugging. In this chapter, we first review the operation of a receiver in high-speed serial interfaces, and especially the clock data recovery block, under the influence of jitter, both in- and out-of-band. Then, we show how a jitter tolerance testing is conducted, including a detailed description of the jitter test signal generation. The key part is presented in a jitter extrapolation technique that will allow us to speed up the tests by inferring about the low BER performance from the actual tests undertaken at higher BER levels, in a fraction of time it would take otherwise.*

Among all the receiver parameters, jitter tolerance is the most challenging to test, validate and debug. Once the jitter tolerance testing is implemented, other tests are either just by-products of the jitter tolerance testing or very straightforward to implement. Jitter tolerance is also the most important parameter in the receiver because of its direct link to the BER performance of the device. This chapter concentrates on accelerating the receiver jitter tolerance testing on ATE. The approaches developed in this chapter are design-independent, and they apply to HSSIs of any type. Non-ATE based testing techniques, such as design for test and loopback-based testing, will be discussed later in Chapter 5.

Jitter tolerance testing for HSSIs requires validating BER performance against the specifications prescribed by various standards at the 10^{-12} BER or lower. This low error rate requirement makes such compliance tests extremely time-consuming. Such a validation normally takes tens of minutes to a few hours at multiple Gbps data rates. As in the ATE world (i.e., in manufacturing test) the test times are measured in milliseconds, it is obviously impractical to adopt this test directly on an ATE.

In this chapter, we demonstrate a technique to perform the jitter tolerance test >1000 times faster. The technique involves extrapolation from the higher BER region down to 10^{-12} BER level for the compliance test. We propose a new model suitable to reason about this extrapolation process. Based on this technique, we then present methodologies to accelerate jitter tolerance characterization and production test.

Y. Fan, Z. Zilic, *Accelerating Test, Validation and Debug of High Speed Serial Interfaces*,
DOI 10.1007/978-90-481-9398-1_3, © Springer Science+Business Media B.V. 2011

3.1 Introduction

3.1.1 Receiver Structure and Characteristics

As discussed in Chapter 2.1, the receiver in the HSSI recovers the transmitted parallel data from the incoming serial stream. Figure 3-1 shows the typical structure of a receiver. The equalizer and the CDR are the two most important blocks that determine the BER performance of the receiver.

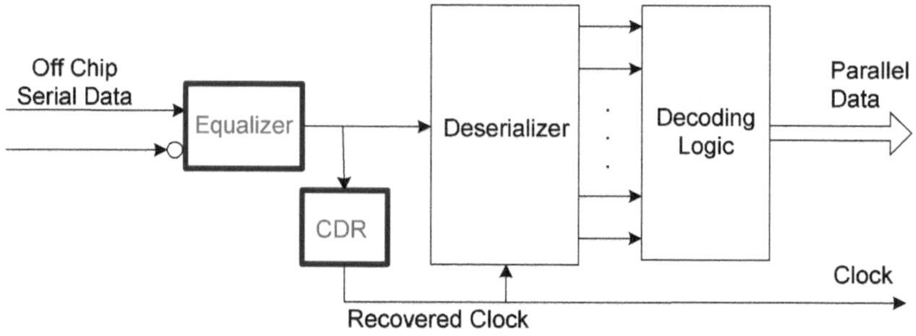

Fig. 3-1. The Typical Structure of a Receiver

The equalizer compensates for the low-pass nature of transmission media, which may introduce ISI at high data rates. According to the cost goal for given applications, different equalization techniques have been proposed. An equalizer can be implemented in either the digital or analog domain, in a linear or non-linear manner, with feed-forward or feedback topologies. Each technique has its advantages and disadvantages. At high data rates, several techniques may be combined together to achieve the best possible performance.

A linear equalizer is relatively simple to implement. However, it is susceptible to noise because it also amplifies the noise while boosting the signal. A linear equalizer hence only addresses ISI but not the crosstalk influence. A non-linear equalizer can in principle amplify the signal but reject the noise. A Feed-Forward Equalizer (FFE) cancels the pre-cursor ISI, where the current symbol is affected by the preceding symbols. A Decision Feedback Equalizer (DFE) cancels the post-cursor ISI, where the current symbol is affected by the post going symbols. An FFE and a DFE are sometimes used together to achieve a design goal. In [143], a serial interface is implemented with a 4-tap FFE, and used in conjunction with a DFE to achieve 6.25/12.5Gbps data rates.

The CDR block is the most critical part of the receiver. It recovers the clock from the received serial data, and then re-times the data using that recovered clock. Figure 3-2 shows a more detailed block diagram of the CDR. The Edge Detector generates all the data edge transitions (both low-to-high and high-to-low transitions). Therefore, the output of the edge detector has the frequency component of the data rate. The edge detector output is then fed into a PLL to recover a clock that is locked to the input serial data rate. The serial data is then re-timed by the recovered clock.

The overall CDR characteristics are mainly determined by the PLL inside the CDR. The structure of a typical linear PLL is outlined in Figure 3-2, within the dash-and-dotted block.

The input signal to the PLL comes from the Edge Detector, which recovers the clock-like waveform from the signal degraded due to a multitude of effects at higher rates. The Phase Frequency Detector (PFD) compares the frequency and phase difference between the edge detector output and the recovered clock signal, and produces narrow control pulses with widths proportional to the phase error. The control pulses are sent to the Loop Filter (LF) to generate a control voltage responsible to adjusting the frequency and phase of the VCO output. The final result is that there is a control system that adjust the VCO to track the input clock rate – it obviously has to be designed to perform such function in a stable and predictable way.

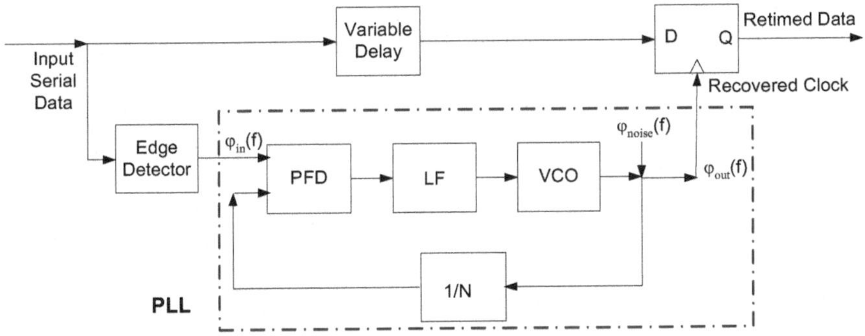

Fig. 3-2. Block diagram of the CDR with a typical linear PLL

To illustrate the characteristics of the PLL, we use $\varphi_{in}(f)$ to denote the PLL input phase of the signal and $\varphi_{out}(f)$ to denote the PLL output phase in Figure 3-2. In the PLL model, $\varphi_{noise}(f)$ represents all of the random VCO phase noise sources, such as the thermal noise and the simultaneous switching noise. We refer the readers to references such as [68] for more details on the noise sources affecting the PLL operation.

Figure 3-3 shows the frequency behavior of the PLL [69]. The function $\varphi_{out}(f)/\varphi_{in}(f)$ is the input jitter transfer function, showing a low-pass characteristic. Therefore, the PLL tracks the jitter in the input signal with jitter frequencies below the PLL bandwidth f_{PLL} (in-band jitter). The function $\varphi_{out}(f)/\varphi_{noise}(f)$ is the VCO jitter transfer function, showing a high-pass characteristic. The PLL will pass through the spectral content of the phase noise that is above the PLL bandwidth (out-of-band jitter).

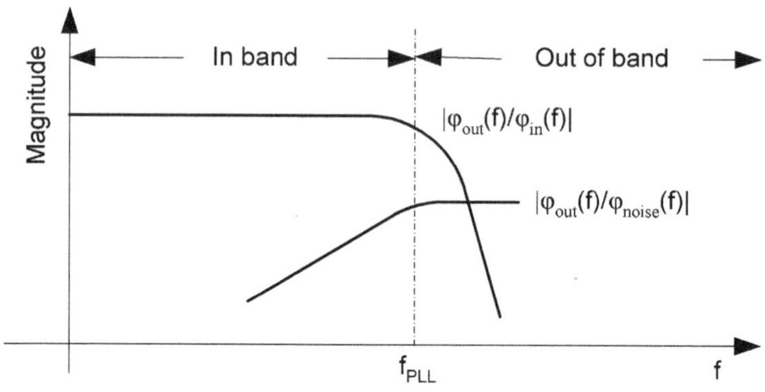

Fig. 3-3. Phase transfer characteristics of the PLL

From the application point of view, the jitter that the receiver sees is observed relative to its recovered clock. Therefore, the output jitter is the timing difference between the recovered clock and the data. Because the PLL has a low-pass transfer function set by the f_{PLL} frequency, the jitter seen by the receiver will have a high-pass transfer characteristic as shown in Figure 3-4, i.e., opposite relative to the input jitter transfer function.

As a rule of thumb often referred to in industry, for the data rate F_d, a typical PLL has a low-pass loop filter with the -3dB frequency at $F_d/1666$, where F_d is the data rate [70]. The jitter transfer function implies that a receiver can tolerate more of the low frequency jitter than the high frequency jitter. Therefore, transmitter jitter and receiver jitter tolerance specifications are defined by jitter frequency bands. The jitter tolerance function is then the mirror image of the jitter transfer function [71], i.e., more jitter can be tolerated when it is known that it will not be transferred further.

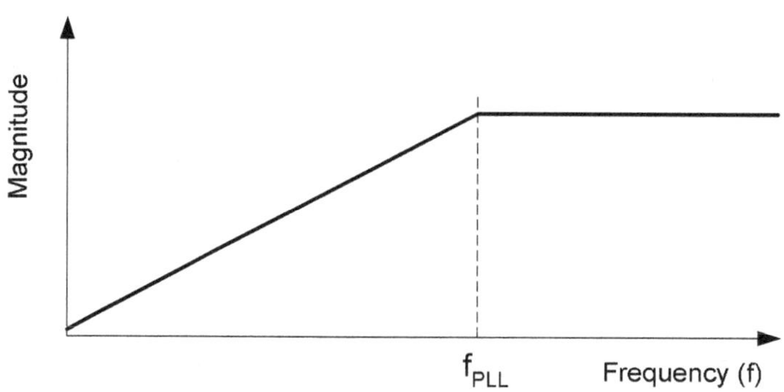

Fig. 3-4. Jitter transfer function of the receiver

Because of this frequency dependence of jitter tolerance, it is important to separate the low frequency and high frequency components of the jitter as usually only the high frequency jitter can degrade the BER performance. Sunter *et al.* propose a technique to separate the high frequency jitter from the low frequency jitter either on-chip or off-chip [44], [72]. The technique uses an undersampling reference clock whose frequency is slightly offset from the primary reference clock. The effective sampling resolution is equal to the difference between the two clock periods. The undersampling technique separates the high frequency jitter from the low frequency jitter based on the unstable regions in aliased samples [44].

Bang-Bang Phase Detector

The above discussion concentrates on the linear PLL. In recent years, the Bang-bang Phase Detector (BBPD) PLL has been gaining popularity in HSSI designs as it is amenabl to HSSI applications. The main difference between a linear and a BBPD CDR is that the BBPD only sends the phase error polarity, early/late, to control the VCO frequency down/up while a linear phase frequency detector outputs a magnitude proportional to the phase error. Compared to the linear PLL, the BBPD PLL has a few unique advantages. First, it does not need to generate narrow pulses and typically can avoid the use of charge pumps [145]. It hence can operate at the highest speed at which a process can make a working flip-flop [73]. In addition, the VCO is undisturbed in the absence of data transitions; therefore it suppresses pattern dependent jitter.

Most modern BBPD PLLs are implemented with some variations of Alexander's phase detector [146]. Two matched flip-flops are usually used: one is driven on the rising edge of the clock while the other is driven on the failing edge of the same clock. When the PLL is locked to data, the two flip-flops sample the incoming data with both the rising edge and falling edge of the VCO clock. Based on three consecutive samples, the phase detector can determine whether the VCO runs faster or slower. Figure 3-5 illustrates the sampling mechanism: after the PLL is locked, the rising edge of the VCO samples the centre of the data bit and produces a retimed data bit (A), and the next VCO rising edge produces a retimed data bit (B). The VCO falling edge between the two rising edges samples the transition (T) between the data bits A and B. Table 3-1 shows the early/late/hold judgments from the three consecutive samples [73]. The VCO frequency is adjusted according to the early/late judgment.

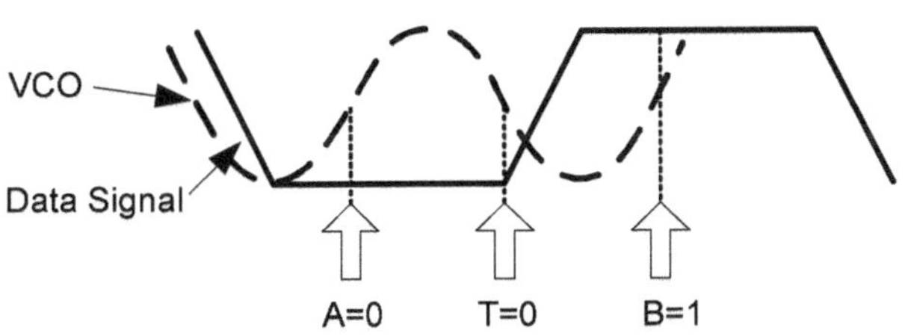

Fig. 3-5. BBPD PLL sampling

Table 3-1. VCO Judgments in BBPD PLL

State	A	T	B	Judgment
0	0	0	0	Hold
1	0	0	1	Early
2	0	1	0	Hold
3	0	1	1	Late
4	1	0	0	Late
5	1	0	1	Hold
6	1	1	0	Early
7	1	1	1	Hold

Although the sampling mechanism and early/late judgment seem to be simple, it is challenging to design well a BBPD and analyze it accurately due to its nonlinear nature.

There are few modeling approaches that have been proposed in literature to facilitate the design and analysis of the BBPD [73], [145], [147]. Figure 3-6 shows the linearized first-order model of a BBPD PLL presented in [73] that we use here as a reference model.

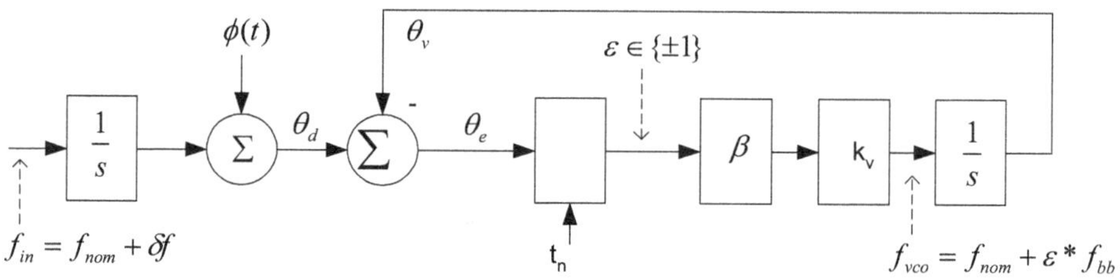

Fig. 3-6. First order model of a BBPD

In this model, $\theta_e(t_n)$ is defined as the difference between the data phase $\theta_d(t_n)$ and the VCO phase $\theta_v(t_n)$ at the nth sampling time t_n. The data phase $\theta_d(t_n)$ can be represented by

$$\theta_d(t_n) = \theta_d(0) + 2\pi \delta f t_n + \phi(t_n)$$

where δf is the frequency difference between the incoming data signal and the VCO centre frequency, and $\Phi(t)$ is the phase jitter with a zero mean.

The phase detector quantizes the loop phase error to a ternary value at each sampling time. The error can be expressed as:

$$\varepsilon_n = sign[\theta_e(t_n)]$$

The error signal ε_n is -1 when the phase is early, 1 when the phase is late or 0 when it is not possible to determine the phase error due to no data transitions. The error signal drives the VCO to produce a change in the frequency of

$$f_{bb} = \beta * k_V$$

where β is an attenuation factor. Therefore, from time t_n to t_n+1, the VCO runs at one of the two frequencies determined by $f_{nom} + \varepsilon_n f_{bb}$ ("hold state discussed later), where f_{nom} is the ideal clock frequency. In a typical CDR, f_{bb} is on the order of 0.1% of f_{nom}. The VCO frequency changes in each cycle. We can approximate the update period by

$$t_{update} = 1 / f_{nom}$$

and the up-or-down phase change (also called bang-bang phase step) by

$$\theta_{bb} = 2\pi (f_{bb} / f_{nom})$$

The loop will remain phase locked as long as the input data signal frequency is in the range of the VCO frequency. Assuming the phase jitter $\Phi(t)$ is small, the input signal frequency error δf needs to be smaller than the f_{bb} frequency, which gives the lock range:

$$-f_{bb} < \delta f < f_{bb}$$

Based on the f_{bb} frequency, we can also calculate the maximum allowed phase jitter. For phase jitter

$$\phi(t) = A \sin(2\pi f_{\mathrm{mod}} t),$$

where f_{mod} is the phase modulation frequency, the instant jitter-induced frequency error is the derivative of the data phase derivation:

$$f_{jitter} = \frac{d[\phi(t)]}{dt} = 2\pi f_{\mathrm{mod}} A \cos(2\pi f_{\mathrm{mod}} t)$$

Assuming δf=0, in order to keep the loop locked, the phase modulation amplitude A at frequency f_{mod} should satisfy

$$A < \frac{|f_{bb}|}{2\pi f_{\mathrm{mod}}}$$

Otherwise, the loop goes into a jitter-induced slew rate limiting. Even though the average input frequency is in the loop lock range, the added jitter causes the instantaneous input frequency deviation to exceed $\pm f_{bb}$, resulting in a transient phase error.

For the first order bang-bang loop, the quantities such as jitter generation, lock range and jitter tolerance are all controlled by one parameter, f_{bb}, which gives us little design freedom. This limitation can be solved by adding another loop to dynamically adjust the VCO center frequency f_{nom} to be equal to the incoming data rate. The second loop can be implemented by an integrator.

We can consider the PLL being composed of two no-interacting branches – an integral branch and a bang-bang branch (or proportional branch). To keep the quality of the first order loop, it is required to keep the phase change due to the proportional branch dominating over the phase change from the integral branch. The stability factor ξ of the PLL is defined as the ratio of two phase changes [73]:

$$\xi = \frac{\Delta\theta_{proportina\,l}}{\Delta\theta_{\text{int}\,egral}} = \frac{2\beta\tau}{t_{update}}$$

The dual branch BBPD structure provides two degrees of freedom: the loop frequency step f_{bb} and the stability factor ξ. In this way, it is possible to make the lock range independent of jitter tolerance and jitter generation. However, more loops also make the PLL design more complicated. To further improve the performance, there is also a trend to implement hybrid phase detectors. The hybrid approach presented in [144] exhibits the intrinsic advantages of low timing jitter in a linear PLL in lock state and the fast locking time in a BBPD PLL.

Overall, there are many factors to consider when implementing a CDR and sometimes we have to make tradeoffs [73], [74], [144] [145]. Nevertheless, as there is a great pressure to characterize and test well the CDR whether it is implemented with a linear PLL or a BBPD PLL, it is desirable to devise testing schemes independent of the design style applied.

3.1.2 Jitter Tolerance Testing Overview

As discussed in Chapter 3.1.1, the CDR is the key differentiator for a serial interface. The overall performance of an HSSI depends on many design choices of the CDR. Due to the economic importance of delivering a correct, liability-free design, we have to perform very thorough characterization testing. This includes the analysis across PVT corners. The traditional jitter tolerance test has always been very challenging. There are mainly two outstanding issues: the long test time, and the complexity to generate a controlled amount of jitter with a proper mix of different types of jitter [34].

The jitter tolerance test is notorious for its long test time. Since most standards for serial links define jitter tolerance performance down to 10^{-12} BER, we need to run 10^{13} bits to check the BER level. Even at 1.5Gbps data rate, it takes 111 minutes (~2 hours) for the device to run so many bits. That is the fundamental limit for running this test fast. With some applications on the trend demanding 10^{-14} BER, the direct measurement is even not practical.

The test time issue becomes much worse when we take into considerations that many design parameters and device settings can affect the jitter tolerance performance. One example is the equalizer settings, which can affect the shape of the waveform sent to the input of the CDR. Another example is the bandwidth of the PLL in the CDR. When the jitter frequency in the input data signal is below the PLL bandwidth, the recovered clock can track the input jitter [69], [75]. However, when the input jitter frequency is above the PLL bandwidth, the jitter gets attenuated more than the lower frequency jitter in the recovered clock. As a result, the jitter tolerance performance is better at lower frequency than that at higher frequency, and the jitter tolerance specifications are defined accordingly to reflect that reality.

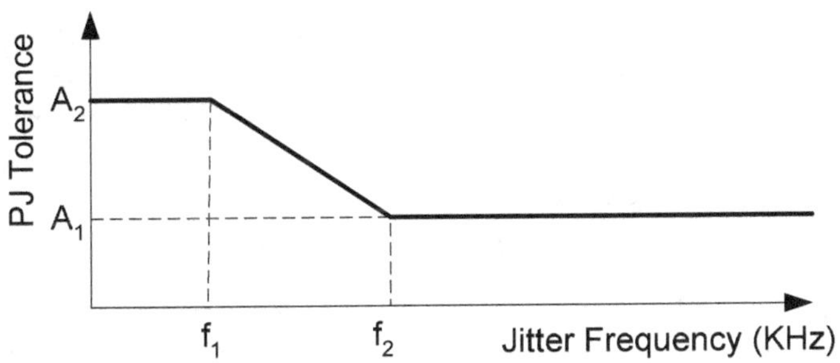

Fig. 3-7. Jitter tolerance specifications

For example, Figure 3-7 shows the jitter tolerance dependence on the jitter frequency for the Serial-attached SCSI (SAS) standard. Here, the defining characteristics are as follows: f_1 is 120KHz, f_2 is 1800KHz, A_1 is 0.1UI and A_2 is 1.5UI [76]. Therefore, the jitter tolerance performance needs to be characterized at multiple frequencies, which can make the jitter tolerance testing extremely time-consuming in practice.

The second challenge for jitter tolerance test is to generate different kinds of jitter and mix them together. For SATA applications, the jitter tolerance test requires a proper amount of deterministic jitter, random jitter and periodical jitter. This is becoming a norm for most of the point-to-point serial links using backplanes or cables [23], [24], [77], [78]. The reason is that the given transmission media generates these specific jitter types.

Jitter is traditionally injected using a few instruments in lab. In [34], a FM source is used to inject PJ; a random noise generator is used to inject RJ; a long cable or a long backplane PCB trace is used to inject DDJ. These kinds of setup can be used to inject large amounts of jitter to test receiver jitter tolerance. However, besides the complexity, it is difficult to mix the different jitter components together and characterize the test signals accurately.

Laquai and Cai propose a jitter tolerance test methodology in [40] based on a DDJ injection filter. This approach uses a passive filter that is carefully tuned to condition the data eye seen by the receiver. This filter can add jitter to stress the receiver. Major advantages of the jitter injection filter are that it takes little space on a loadboard and that its cost is very low. However, this methodology does not offer the flexibility of varying the amount of injected jitter. In addition, the amount of the injected jitter is very sensitive to the data rate. For example, for the same filter and the same transceiver, the injected jitter changes from 0.25UI at 2.125Gbps to 0.42UI at 2.67Gbps [40]. This limits the applications of that approach, especially that it has become quite common for current HSSIs to accommodate multiple protocols and hence require the test equipment to have the ability to generate a controllable amount of jitter.

Hafed *et al.* propose a high density HSSI tester that can significantly reduce the HSSI validation time by relying on parallelism and efficient measurement techniques [41]. In this tester, the jitter is injected by modulating the PLL input signal of the transmit port module. The advantage of this approach is that the high-speed signal path is completely untouched. Hence the approach would not produce unwanted jitter in the high-speed signal path when the injected jitter is programmed to zero. However, the injected jitter frequency is limited to the tracking bandwidth of the PLL. Chapter 3.5.1 will demonstrate that any jitter with frequencies above the PLL bandwidth will be drastically attenuated.

Keezer *et al.* present another modular approach in [42] for testing multiple Gbps HSSIs. Their approach eliminates the PLL bandwidth limitation by modulating the high-speed clock (PLL output signal) that has the same rate as the Gigabit serial data signal. The jitter is injected to the clock signal through dynamically shifting its phase by an ATE device. Advantest has also developed a jitter injection module that can mix any modulated signal as a jitter signal with a carrier signal [43].

Sunter *et al.* propose an alternative approach in [44] for the jitter tolerance testing. Instead of injecting jitter to the test signal, their approach uses on-chip undersampling to measure parameters that affect the jitter tolerance, such as high-frequency jitter in the received signal and the recovered clock. This approach requires that the CDR be selected to lock either to the received serial data or to an offset reference clock, and that the parallel data ports be fully accessible. A module called UnLimited Time Resolution Analysis (ULTRA) is used to extract all the measurements. The ULTRA module can be placed either on the loadboard or on-chip. However, adding an extra module onto a loadboard or a device requires extra space and cost that in many cases cannot be tolerated. It also complicates the design and test program development.

On ATE, supplying the proper mix and amount of jitter is always more challenging. Even though there are existing jitter injection techniques that are targeting the test of HSSIs [41], [42], [44], they usually need extra add-on modules on the testing loadboard or inside the device as a Design-for-Test (DFT) feature. To actually implement them in an ATE environment needs a lot of considerations, such as the loadboard design complexity and the impact to the overall design cycle time. In this chapter, we present a jitter injection technique based on an Arbitrary Waveform Generator (AWG) on ATE, which does not need any extra module or add-on circuits. This approach can qualify the jitter tolerance performance of HSSIs with data rates up to 3Gbps. Another jitter injection technique based on the state-of-the-art phase delay product will be presented in Chapter 5, which can be used to test HSSIs with data rates up to 12.5Gbps.

3.1.3 Proposed New Method

To address the two outstanding issues in jitter tolerance testing, we aim to develop schemes that can perform jitter tolerance testing on ATE in a much faster manner. ATE is well known for its high throughput in production. ATE-based solutions are also more widely being used in validation and characterization, especially for performance analysis across process, voltage and temperature corners on a large sample size.

In this chapter, we propose a new approach that can perform the jitter tolerance test >1000 times faster. The approach is straightforward: we will be varying the amount of input jitter to get the receiver into several higher BER levels, from which we will then extrapolate down to the 10^{-12} BER specification for jitter tolerance qualification. The conceptual illustration is shown in Figure 3-8. We will present the concept, assumptions, extrapolation models and experimental data in the rest of this chapter

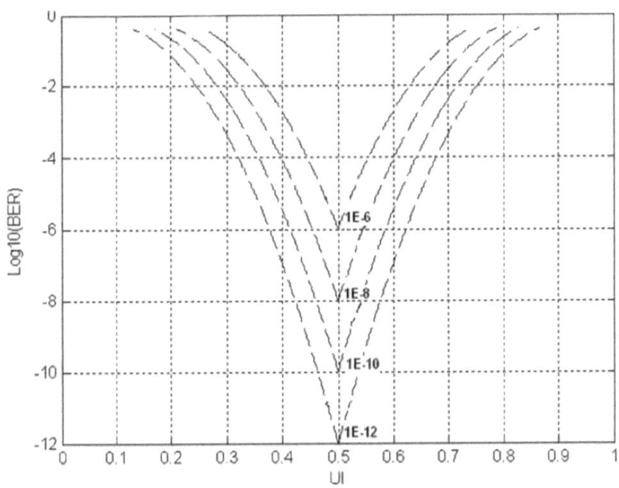

Fig. 3-8. Conceptual illustration of the jitter tolerance extrapolation

Jitter tolerance extrapolation is a relatively new subject in HSSI testing and validation. We start by transplanting the main ideas from transmitter jitter measurements. The histogram tail-fitting, real-time sampling and Time Interval Analysis (TIA) approaches require the knowledge of the actual probability density function of the jitter. For the receiver CDR, that information is not as accessible as in the transmitter. Therefore, we cannot use the jitter PDF for receiver testing. Only the BER scan (also known as bathtub curve) uses a model based on certain assumptions, which we may use for receiver jitter tolerance testing.

In the BER-scan based transmitter jitter measurement, a bathtub curve is usually used. The bathtub curve is generated through sweeping the sampling position on the timing axis and then recording BER at each sampling position. Figure 3-9 shows an example of the bathtub curve. It is actually a plot of data eye openings at various BER thresholds. The finite slope of the bathtub curve is caused by RJ. Obviously, at a lower BER, the eye opening becomes narrower. Figure 3-9 also shows a simplified formula to calculate DJ, RJ and TJ with only two data points recommended by the XAUI standard [34].

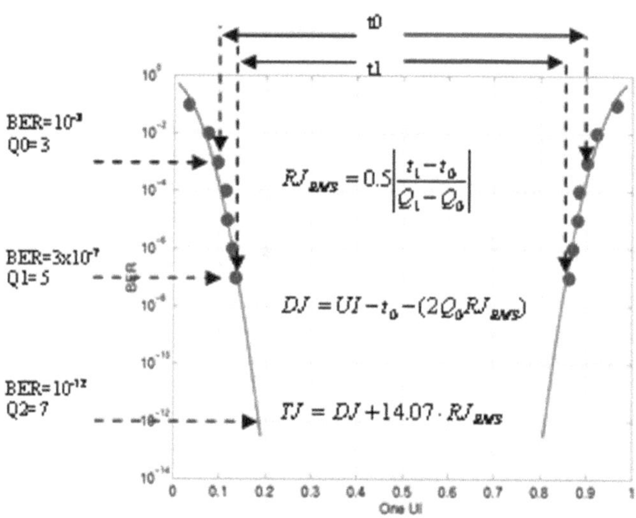

Fig. 3-9. Transmitter BER scan

In the bathtub based transmitter jitter measurement, all that we can observe is the combined TJ profile, which is a convolution of the DJ and RJ. We need to work backwards to derive its DJ and RJ components. This separation is nearly impossible if we do not make some assumptions and use a simple enough model to facilitate the analysis.

The most popular model used is the so called "double delta" model – assuming that the only DJ is DCD, whose PDF is only comprised of a pair of delta functions [80]. In this case, the complicated convolution is reduced to a standard complementary error function. The double delta assumption now serves widely as the model for modern bathtub curve fitting.

Theoretically, the double delta model is not the most flexible one for arbitrary jitter profiles. Studies on the effect of the DJ profile when deviating from the double delta assumption show that there are limitations of the double delta model [80]. However, this seemingly limited model works reasonably well when used appropriately (i.e. with the proper selection of the curve fitting range). This particular model is favored by many engineers because it directly links the jitter to the system-level BER performance. Further, the eye openings at lower BER levels can be extrapolated from the openings at high BER levels. The extrapolation results can be verified by performing direct measurements at the lower BER levels. Therefore, we show here how to borrow ideas from the transmitter BER scan for our jitter tolerance extrapolation.

Similar to the transmitter BER scan, we perform a receiver BER scan. We collect data points at higher BER levels, which takes much less time to do. Then we extrapolate performance to the lower BER range. The extrapolation accuracy can be easily verified by performing the test at the lower BER range, and comparing it to the extrapolation result. If the error is small, then the new method is considered acceptable.

However, there are two significant differences between the receiver BER scan and the transmitter BER scan. First, the receiver BER scan is no longer going to be a bathtub curve fit. We usually cannot control the data sampling position in a receiver because the CDR circuitry is designed to sample the data in the middle of each bit.

Therefore, the receiver always works at the crossing point of the bathtub curve. The BER of a receiver is associated with the jitter in its input signal. As shown in Figure 3-10, with the increase of the jitter in the input signal, the bathtub curve moves up, indicating a higher BER. What we get is a series of crossed curves, as shown in Figure 3-8. Therefore, a new extrapolation algorithm needs to be developed for the jitter tolerance test.

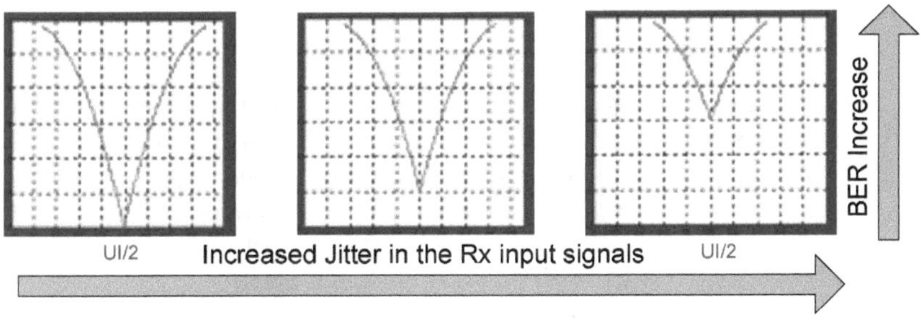

Fig. 3-10. Receiver BER scan

Another significant difference is that in the receiver BER scan we have the control over the jitter PDF of the test signal, while on the other hand, we do not have that level of control over the jitter PDF in the transmitter. Normally, the HSSI standards define separate DJ, PJ and RJ specifications, but they do not define the shape of the jitter probability distribution. In the jitter tolerance test, it is important to note that we have some flexibility to shape the jitter PDF. This is a critical property because different types of jitter profiles can significantly affect the curve-fitting accuracy when we employ the complementary error function to model the curve [80].

Considering that in the transmitter BER scan, the jitter profile with a double delta PDF gives a better curve fit, we choose to inject single-tone sine wave jitter to the receiver test signal, which would generate a jitter profile similar to a double delta distribution.

As shown in Figure 3-11, there is also a strong tendency to favor the two edges in the sine wave PDF curve. Therefore, it would be a closer fit to the complementary error function. Another reason to inject the sine wave jitter is that only sinusoidal jitter can provide the worst-case jitter to HSSI devices as elaborated in [81].

Sinusoidal jitter is commonly used to perform jitter tolerance testing [82], [83]. Even though in the receiver input signal there are other DJ components (e.g. ISI) that may change the jitter profile a bit, we are not concerned about them. In the transmitter jitter bathtub curve fitting, we can still achieve good accuracy with many kinds of jitter distribution.

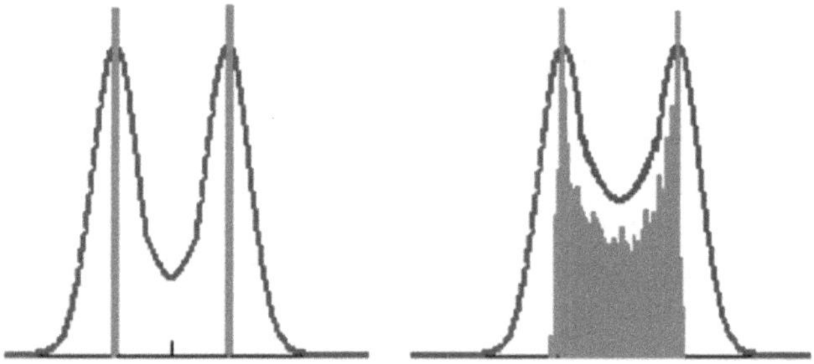

(a) Double delta PDF (b) Sine wave PDF

Fig. 3-11. Jitter PDFs for curve fitting

In this chapter, we propose to inject the sinusoidal PJ using an AWG available on ATE. We can generate controllable amounts of PJ with only one piece of equipment – AWG. In the receiver BER scan for jitter tolerance testing, the AWG output is directly connected to the input of the receiver as shown in Figure 3-12. By varying the amount of injected PJ, we can get different BER data points.

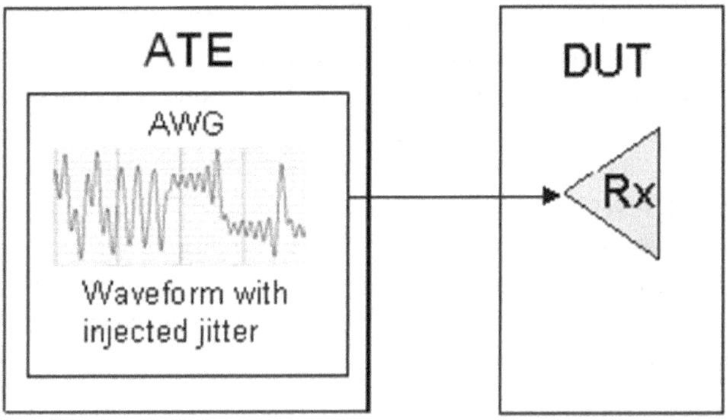

Fig. 3-12. Test setup for jitter tolerance testing

The test setup of the proposed solution is very simple. There is no intermediate add-on circuit, which means that we do not have to switch off any circuit for the functional test and the input sensitivity test where clean signals (no jitter is injected) are used. The AWG approach can also produce high frequency PJ, like the more modern Voltage-Controlled Delay Line (VCDL) modulators. The high frequency PJ is needed for testing CDR out-of-band jitter tolerance where the CDR can no longer track the input PJ.

In the remainder of this chapter, we will present the details on how the test signal is generated, how the bit errors are detected, how we develop an algorithm for jitter tolerance extrapolation and, finally, how we use the algorithm to accelerate jitter tolerance qualifications.

3.2 Jitter Test Signal Generation

On ATE, there are mainly two types of instruments that can perform GHz receiver testing: AWGs and binary digital pattern generators. Generally, binary digital pattern generators such as Teradyne SPQ [84] and Agilent/Verigy NP3G [85], can provide higher analog port count, suitable for multiple serial interface testing. However, an AWG-based approach provides us more controllability to the signal it generates, including jitter injection and multiple-level amplitude manipulation.

In addition, AWG-based solutions exhibit more capabilities in testing other analog blocks: the same instrument can be used to provide test signals for unrelated blocks, such as an amplifier and ADC. This is especially attractive for SoCs, which have a very limited number of HSSIs (typically 1 or 2), but have many other analog blocks. AWG-based approaches can provide better overall cost efficiency in this application. This chapter concentrates on the AWG-based approach. Chapter 5.3 proposes a test solution suitable for multiple-port HSSIs, such as networking and switching devices.

To perform a jitter tolerance test, we need source signals with controllable jitter. In our implementation, the source signals are generated by modulating the ideal AWG binary signals with a user defined jitter profile. Generally speaking, we can source any waveform with spectral contents limited to the Nyquist band. The jitter injection approach does not require additional instruments as in some other setups [78], [83].

Using the state-of-the-art AWG on ATE (6G samples/s and 2.0GHz analog bandwidth), we have 2 samples per bit for 3 Gbps data. By manipulating the sample timing and amplitude, we can inject a controllable amount of jitter to the test signals. Using the AWG6000, we can perform the receiver jitter tolerance testing for data rates up to 3Gbps and the receiver function verification for data rates up to 6Gbps.

The main principle of our jitter injection mechanism is to modulate jitter-free data edges using a jitter signal. Oversampling, FFT and downsampling techniques are further used to achieve high-quality test signals with a desired jitter resolution. The generated signals are essential to characterize the jitter tolerance performance along with other receiver parameters. The following section describes the details of the jitter injection scheme.

3.2.1 Choosing Test Signal Parameters

In our implementation, AWG6000 is used and its sample rate is set to 6GHz. Each data bit has two samples for 3Gbps signals and four samples for 1.5Gbps signals. The AWG6000 can also be used to generate test signals with non-integer samples per bit, such as for the 5.5Gbps application discussed in Chapter 3.2.3. The following jitter injection discussion is based on certain data rates for illustration purposes; the principle is the same for other data rates, but minor changes are needed based on specific requirements.

To generate test signals with controllable amount of jitter for jitter tolerance testing, we need to properly choose or set the following parameters:

- Test pattern
- Length of the test signal
- Jitter signal
- Jitter injection resolution

As for the test pattern, it should be able to represent the bit patterns occurring in real applications. Pseudo Random Bit Sequence (PRBS) patterns have been widely accepted as a means to test different communication interfaces because they have different run lengths and hence provide very good test coverage for possible data combinations.

For instance, in SATA applications, the maximum run-length is 7. Therefore, for this class of applications we choose the standard 128-PRBS pattern, which has a maximum run-length of 7. Figure 3-13 shows the contents of the 128-PRBS pattern. The 128-PRBS covers all run-lengths from 1 to 7. There are totally 63 transition edges in the test pattern. We inject jitter to the PRBS signal through modulating these transition edges. The injected jitter can have any arbitrary profile. In this section, we demonstrate how different amounts of sinusoidal jitter are injected to the data signal.

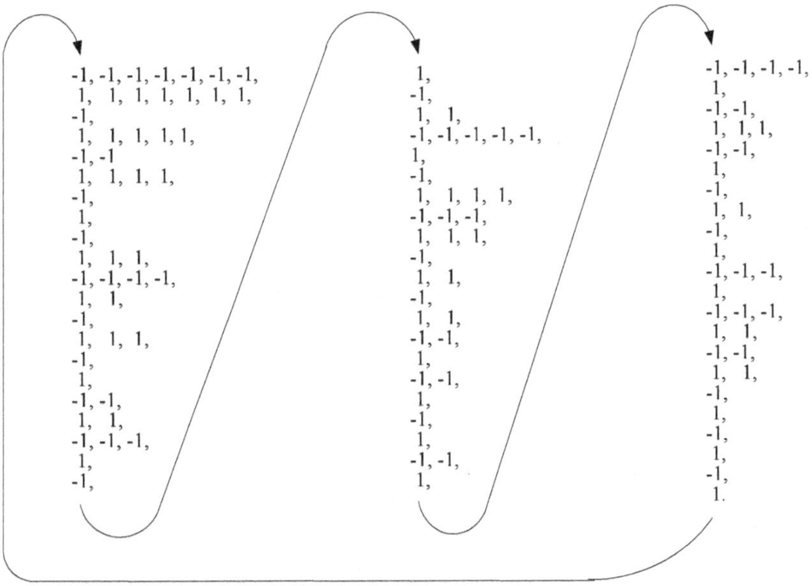

Fig. 3-13. The 128-PRBS sequence

Since the length of an AWG sequence is finite, the test signal (including data and jitter) is generated by running the AWG pattern continuously. The injected jitter signal needs to be coherent with the data pattern stored in the AWG. In [2], we modulate one cycle of the 128-bit PRBS data pattern using a sinusoidal signal. The generated test signal exhibits the exact amount of PJ as we expected according to the calibration result using standalone lab equipment.

However, with the test signal, ATE reports better jitter tolerance performance than that observed on bench equipment when we characterize some special devices using the same 128-bit PRBS pattern. Further experiments demonstrate that the difference is caused by different jitter injection mechanisms between the bench and the ATE. On bench, the PJ signal and the un-modulated data signal are asynchronous -- each edge of the data signal can be modulated up to 100% of the injected PJ peak value. On ATE, the jitter and the un-modulated data signal are always synchronous: the jitter at each of the 63 edges of the 128-bit PRBS pattern is a constant. Therefore, not every edge has a chance to be modulated with the peak value of the injected PJ signal -- some edges even do not move.

We overcome this issue by increasing the length of the ATE test pattern. If the movement of each edge of the test signal can reach or nearly reach the PJ peak value, the bench test signal is emulated. For this reason, along with the ATE memory source availability and FFT requirements, we increase the length of the test pattern by repeating the 128-bit PRBS pattern 2^n times, where n is an integer. We then modulate the long test pattern using a sinusoidal jitter signal with an odd number of cycles. This can greatly increase the randomness of the edge movement for each edge of 128-bit PRBS pattern.

Our calculation shows that a good choice is to use a 1024-bit test pattern (constructed by repeating the 128-bit PRBS pattern eight times) and then to modulate the test pattern using a sinusoid PJ signal with 39 cycles. In the generated test signal, each edge of the 128-bit PRBS can reach at least 92% of the injected PJ peak value, which is very close to the bench test signal. Even though further increasing the length of the test pattern can slightly further increase the randomness of the edge modulation, choosing 1024 bits is a good tradeoff between the randomness and the AWG memory usage (we need to store multiple test signals in the AWG for jitter tolerance characterization and other tests).

When 1024-bit test signals are used in 3Gbps applications, the FFT frequency resolution is given by

$$f_{awg\_FFT\_res} = \frac{3000MHz}{1024} = 2.9296875MHz$$

We can inject any sinusoidal jitter signal that is a multiple of $f_{awg\_FFT\_res}$. Lower frequency jitter can be injected by increasing the length of the test pattern. In our experiments, we investigate the jitter tolerance characteristics at different frequencies while concentrating on the jitter frequency of 114.2578125MHz (= 39 x 2.9296875MHz) for most of our work. At this frequency, the generated test signal exhibits very good randomness of edge modulation. This frequency is also a good representative of the out-of-band frequency in SATA, whose range is from 6MHz to 300MHz [22].

3.2.2 Periodic Jitter Injection

We inject jitter to the jitter-free data signal by modulating the ideal data edges -- moving them forward when the jitter is positive or backwards when the jitter is negative.

The process of jitter injection consists of the five steps outlined in the following five subsections.

3.2.2.1 Creating Jitter-Free Data Signal

The jitter resolution of the modulated data signal is directly determined by the time resolution of the data edges. For instruments such as AWG6000, the maximum sampling rate is 6G samples/s. Each data bit can have two samples for 3Gbps signals and four samples for 1.5Gbps signals. If we directly manipulate the edge transitions, the jitter resolution is only 0.25UI even with the 1.5Gbps signals because each AWG sample represents 0.25UI. We cannot use this approach to perform jitter tolerance testing because pico-second jitter resolution is required.

One solution is to oversample the jitter-free data signal to pico-second resolution. Considering the requirement of FFT that we will perform later, we need to choose the oversampling rate to be a power of 2. For the 1.5Gbps signal, one UI is 667ps. We can choose an oversampling rate of 512 samples per bit, which translates into a jitter resolution of 1.3ps. Figure 3-14 shows one cycle waveform of the 128-bit PRBS jitter free data signal *oversampled_data* oversampled with 512 samples per bit.

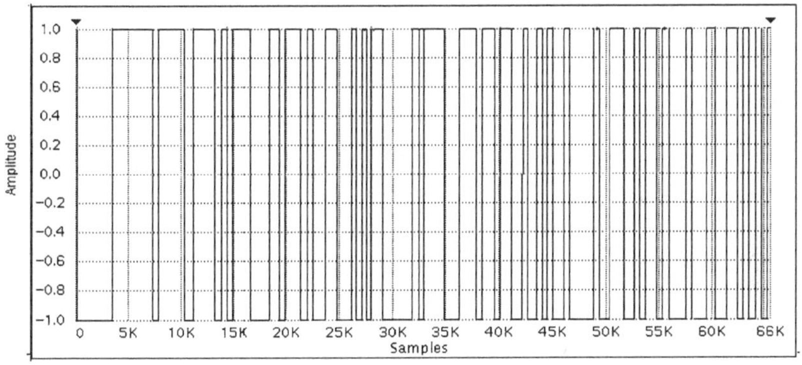

Fig. 3-14. Oversampled jitter-free data signal

3.2.2.2 Creating a Digitized Jitter Signal

Now we have created the jitter-free data signal. Next, we need to modulate the data edges in order to convert jitter amplitude information to timing information. This is done by moving the edge of the data signal based on the jitter amplitude information. Therefore, we need to create a digitized PJ signal to modulate the ideal data edges.

A sinusoidal jitter signal can be characterized by two parameters: the frequency and the amplitude. The frequency can be represented by $pj\_bin * f_{awg\_FFT\_res}$, where *pj_bin* is the PJ frequency bin. The amplitude is the PJ peak value *amp_UI*, which is the maximum edge displacement in the modulated data signal. Based on the jitter parameters, the digitized jitter signal *jitter[i]* can be represented by

$$jitter[i] = amp\_UI*\sin(2*M\_PI*i*pj\_bin/1024 + M\_PI/2)$$

where i is the sample index and $i \in [0, 1023]$, *amp_UI* is the jitter amplitude measured in UI, *pj_bin* is the jitter frequency bin and *M_PI* is the constant 3.14159....

The jitter amplitude is measured in UI. Therefore the value of *jitter*[*i*] represents the data edge displacement in UI at data bit *i*. As the data signal is oversampled with 512 samples per bit, the data edge timing resolution is UI/512. To convert the jitter into data edge time displacement, we need to oversample the *jitter*[*i*] with a resolution of UI/512, which gives

$$oversampled\_jitter[i] = 512 * jitter[i]$$

where *oversampled_jitter* [*i*] represents the number of samples that the data edge needs to be pushed back or forwards.

For the case of 1.5Gbps signal, one UI is 667ps and each data bit is oversampled by factor of 512. If we set the jitter peak-to-peak value to 400ps, the jitter amplitude *amp_UI* is 0.3UI, *jitter*[*i*] is between 0.3 and -0.3, and *oversampled_jitter*[*i*] is between 153 and -153.

3.2.2.3 Modulating the Data Signal

Each bit of the oversampled data is then modulated by the oversampled jitter signal. Depending on the jitter amplitude and polarity at each data bit, the data transition edge is pushed forward or back by manipulating the *oversampled_data* samples. The following pseudocode demonstrates how the modulation algorithm is executed:

```
for (i=0; i<=1024; i++)
{
   if oversampled_jitter[i] > 0
   {oversampled_data[i*512]~oversampled_data[i*512+oversampled_jitter[i]]
      is replaced by oversampled_data[(i-1)*512]     //push the edge later
   }
   else
   {oversampled_data[i*512-oversampled_jitter[i]]~oversampled_data[i*512]
      is replaced by oversampled_data[i*512]         //pull the edge earlier
   }
}
```

Figure 3-15 illustrates the first 32 bits of the data signal and the jitter signal. The data signal has been oversampled by 512 and modulated by the jitter signal. At the first transition edge (bit 7), the jitter is negative, so we pull the transition edge earlier; at the second edge, the jitter is positive, so we push the transition edge later; for the fourth edge, the jitter is 0, so the transition edge does not change.

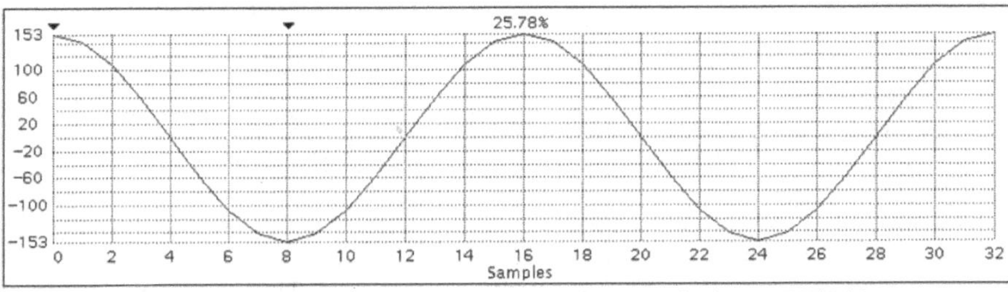

(a). Jitter signal *oversampled_jitter*: Vpp = 0.6 UI, resolution = UI/512

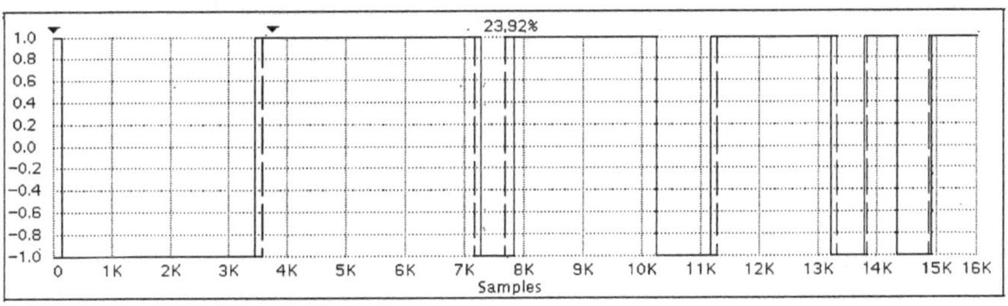

(b). Data signal: jitter-free data (broken line) and jittered-data (solid line)

Fig. 3-15. Jitter signal and modulated data signal

3.2.2.4 Generating Bandwidth Limited Signals

The above oversampled jittered data signal is an ideal signal that has zero transition time: it contains infinite frequency spectrum and does not suffer from any bandwidth limitation. However, the actual AWG we use only has a bandwidth around 2G, and its maximum sampling rate is 6G samples per second. In order to retain the jitter and maximize the signal to noise ratio, we use two techniques: the first one aims to smoothen the transition edges and second one limits the bandwidth of the signal.

To smoothen the transition edges, we add a transition time. In this example, we set the transition time from rail to rail to be 0.8UI (410 samples). At this step, the transition is linear. Band-limited filtering in the following step will generate the signal with "real" transitions. Figure 3-16 shows the data signal with linear transitions.

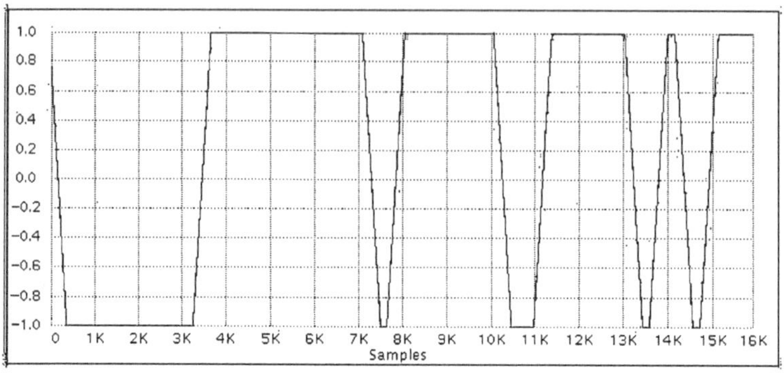

Fig. 3-16. Adding edge transition time

To limit the modulated signal bandwidth, we need first to obtain the magnitude frequency response and phase response of the data signal. This can be achieved by performing FFT on the modulated oversampled data samples.

Figure 3-17 shows the frequency spectrum of the modulated data signal. Here, only the first 160 bins of the total of 262144 bins spanning the frequency band are displayed.

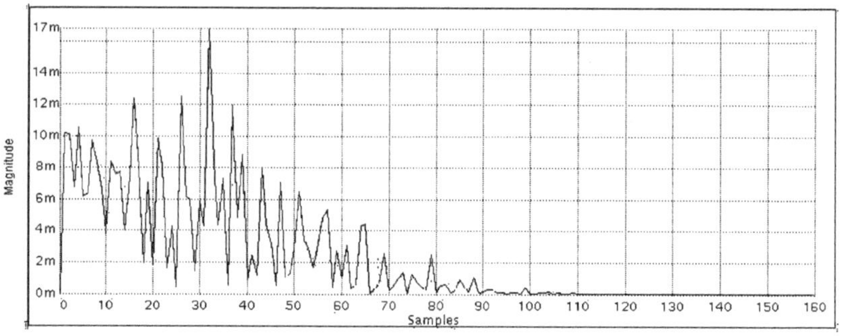

Fig. 3-17. Frequency spectrum of the oversampled data signal

In this case, we limit the bandwidth of the generated test signal to 3GHz because the AWG sampling rate is 6GHz. Therefore we filter out all frequency components above the Nyquist frequency by setting these frequency components to zero.

By performing an inverse FFT operation, we can get the time domain data. Figure 3-18 shows the waveform of one cycle of the 128-prbs pattern after the inverse FFT operation.

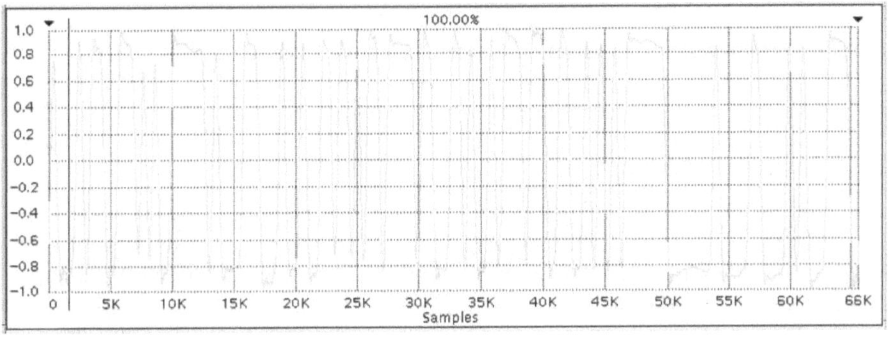

Fig. 3-18. Time domain data after the inverse FFT

3.2.2.5 Downsampling to Get AWG Samples

The data after the above manipulation is still oversampled with 512 samples per bit. If we directly store the oversampled data in the AWG for 1.5Gbps application, the AWG sampling rate needs to be as high as 768GHz. Because the AWG sampling rate is limited to 6GHz, we need to decimate the waveform to get the desired AWG samples.

For 1.5Gbps applications, we need to keep 4 samples out of 512 samples (one data bit), which translates into a downsampling rate of 128. Figure 3-19 shows the waveform of 128 bits of the final data signal that we generate, where each bit has been represented by 4 samples. This data can be directly stored in the AWG memory.

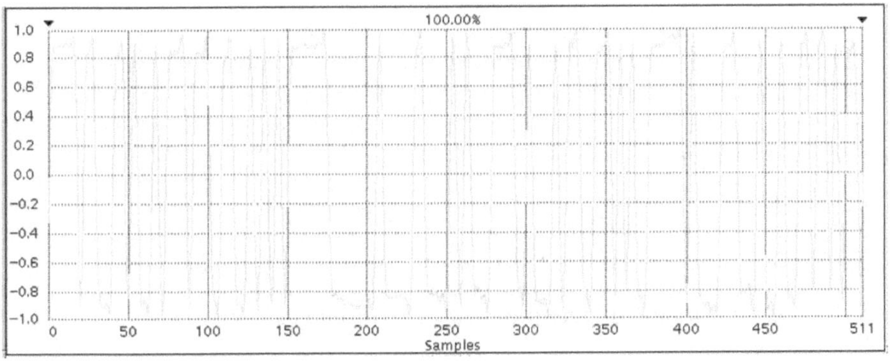

Fig. 3-19. AWG waveform – data after under-sampling

With four samples per bit, the 1024-bit test signal takes 4k AWG sample source memory. Because the AWG6000 has 32M sample source memory, we can use one AWG to store multiple jittered test signals, which are needed when we sweep the test signals to characterize the jitter tolerance performance. The same

AWG can still be used to store other waveforms to test other blocks in an SoC. In addition, we can test multiple HSSIs in parallel if we have multiple AWGs.

3.2.3 Fractional Sampling

The test signal generation scheme discussed previously is based on integer sampling – each data bit has an integer number of samples in the AWG6000, such as 2 samples per bit for 3Gbps signals and 4 samples per bit for 1.5Gbps signals when the sampling rate is set at 6G samples per second. Normally there is only a limited range of sampling frequencies that we can set for the AWG in order to optimize the AWG performance. For example, for the AWG6000 waveform generator, the sampling frequency can only be set to rates between 5.8GHz~6.2GHz or below 5GHz.

If the sampling rate cannot be set to a multiple of the data rate we need to investigate, fractional sampling has to be used - each data bit has a non-integer number of samples. For example, if we need to investigate a 2.75Gbps application, we need to use fractional sampling because the AWG sampling rate cannot be set to 5.5G. This section presents a method to generate test signals with fractional sampling. The following demonstrates how the test signals for the 2.75Gbps application are generated using the 6G sampling rate.

When the AWG sampling rate f_{awg} is set to 6G and the input data rate f_{in} is 2.75Gbps, each data bit needs to be sampled by

$$\frac{f_{AWG}}{f_{in}} = \frac{6.0}{2.75} = \frac{24}{11} \tag{3-10}$$

which shows that every 11 data bits need 24 AWG samples. Therefore, the length of the data pattern needs to be a multiple of 11. For example, if we want to use a 160-bit test pattern, we need to repeat the 160-bit pattern 11 times, which means we need to increase the test pattern to 1760 bits and the number of AWG samples should be 3840 according to

$$\frac{f_{AWG}}{f_{in}} = \frac{24}{11} = \frac{24*160}{11*160} = \frac{3840}{1760}$$

To convert 1760 bits of data into 3840 AWG samples, we propose the following procedure:

1) Oversampling the data by a multiple of 24 (derived from Equation (3-10)). We can choose an oversampling rate of 240. This oversampling translates into a resolution around 1.5ps for the 2.75Gbps data signal

2) Performing an FFT on the 1760 bits data oversampled by 240 (the total number of samples is 422400) to covert the time-domain data into frequency domain data
3) Modulating the data edge using a jitter signal if needed using the procedure discussed in Chapter 3.2.2
4) Filtering out all the frequency components above 3GHz by setting these frequency bins to zero
5) Performing an inverse FFT to convert the band-limited frequency domain data into time-domain data
6) Downsampling the time-domain data by 110 to get the 3840 AWG samples

If we run the AWG with the generated AWG samples continuously and set the sampling frequency to 6G, the AWG would generate a 2.75Gbps data signal as we need. It is equivalent to that each data bit has 2 and $^{2}/_{11}$ AWG samples. For other data rate applications, we only need to adjust the oversampling and downsampling ratio based on f_{AWG} and f_{in} when applying the above procedure.

3.2.4 Jitter Calibration

We need to thoroughly calibrate the generated test signals for two reasons. The first one is to verify our jitter injection technique. Secondly, we use the calibration results to link the injected PJ to DJ and TJ because most HSSI standards define DJ and TJ tolerance specifications separately and we control the DJ and TJ through the injected PJ.

The jitter injection technique is verified by extracting the actual jitter in the generated signal and then comparing it with the injected value. There are three approaches we can use for the verification. The quickest approach is to plot the eye diagram according to the test signal data stored in the AWG. Another approach is to use a digitizer on the ATE to capture the AWG output signal and then extract the jitter information. The jitter in the generated test signal can further be calibrated using bench equipment.

We first verify our jitter injection technique using eye diagrams. Figure 3-20 captures the eye diagram of a 1.5Gbps test signal with 300ps PJ injected. The diagram is generated by overlaying the data samples stored in the AWG in one UI interval, and the figure is plotted in MATLAB. As we can see, the eye closure is very close to the amount of PJ we injected. Without the need for any instrument, the eye-diagram can be generated and used to verify the concept of our jitter injection technique.

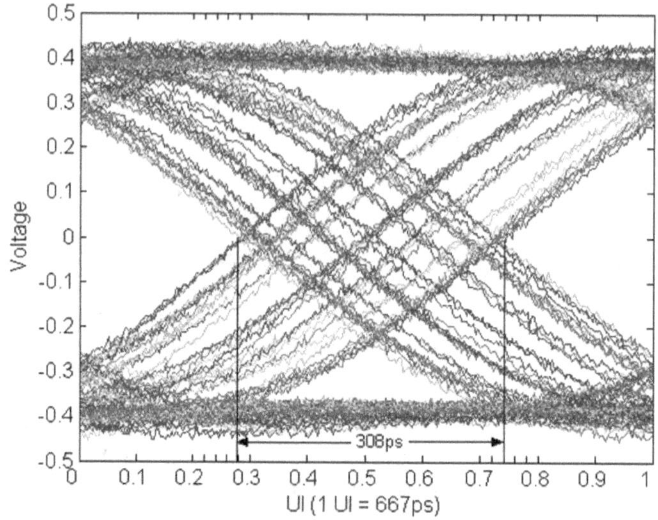

Fig. 3-20. The eye diagram of the AWG samples with 300ps PJ injected

We can calibrate the amount of injected jitter on ATE by connecting the AWG output to the input of a high bandwidth digitizer, such as the GigaDig, available on systems such as Teradyne Catalyst/Tiger ATE. Figure 3-21 shows the connection used to calibrate the injected jitter on ATE. We use the digitizer to capture the AWG output and we then extract the jitter components from the captured waveform [4]. The details on how the jitter components are extracted are furthers discussed in Chapter 4.

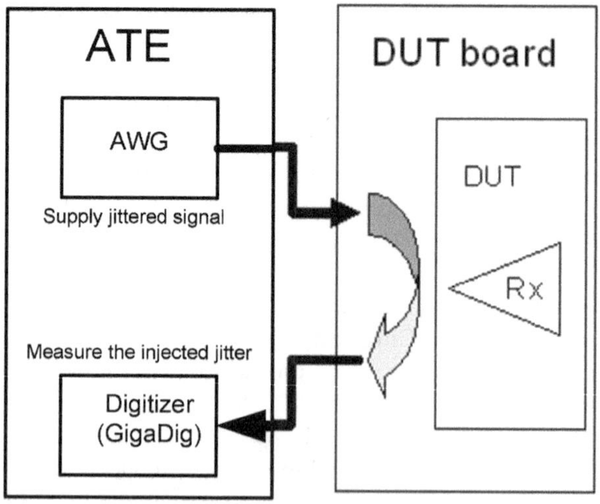

Fig. 3-21. Test setup for jitter calibration on ATE

Figure 3-22 plots the jitter calibration result using the ATE. The horizontal axis is the injected PJ and the vertical axis is the measured jitter. As we can see, the measured PJ correlates well with the ideal injected PJ. We have a good linear control on the injected jitter. DJ and TJ are also recorded. There is a small offset between the injected PJ and the measured DJ, which is contributed by other DJ components. The offset between the measured DJ and the measured TJ is caused by intrinsic RJ of the AWG.

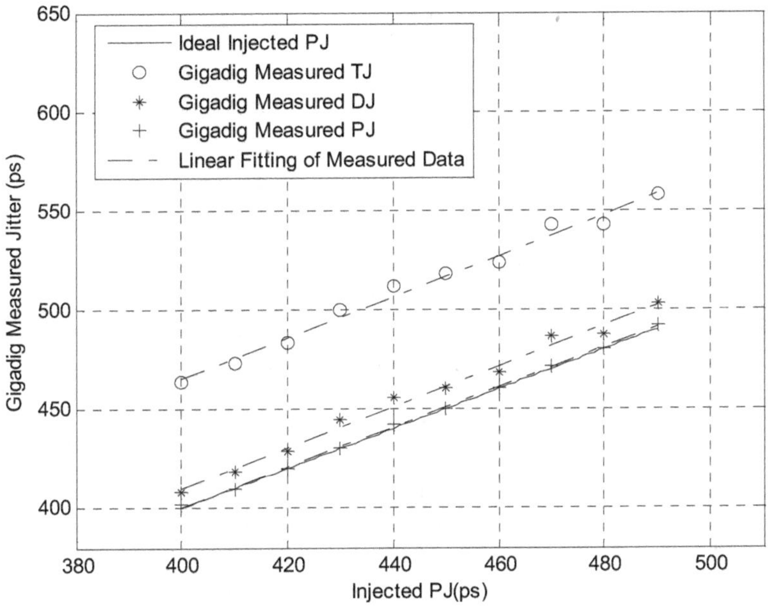

Fig. 3-22. Jitter injection calibration curves with the digitizer

To further confidently report the jitter numbers, we used a Wavecrest SIA-3000 to calibrate the TJ, RJ and DJ numbers in the generated test signals. Figure 3-23 plots the measurement results. Similar to the ATE jitter measurement results shown in Figure 3-22, the Wavecrest measured DJ and TJ are tracking the injected PJ. The constant vertical offset between the measured TJ and the injected PJ is caused by the intrinsic RJ, DCD and ISI from the AWG and the connection cables. In this example, the offset between the PJ and TJ is around 92 ps.

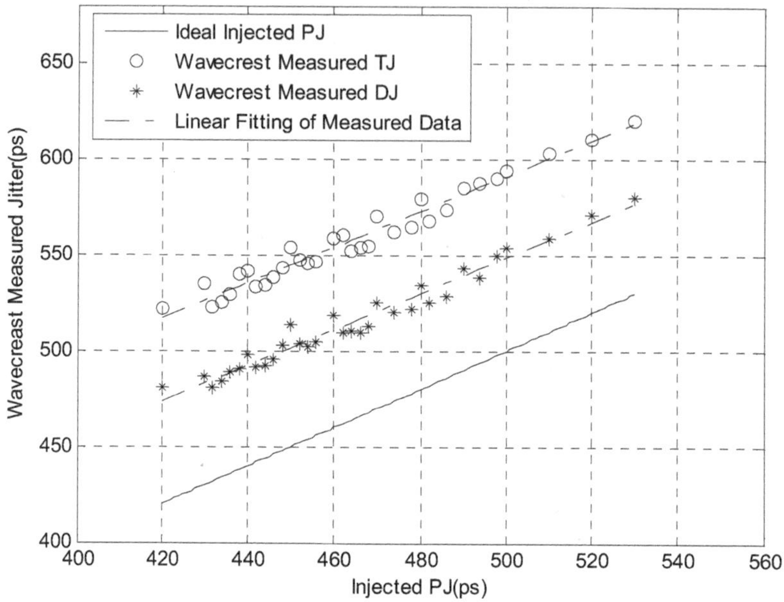

Fig. 3-23. Jitter injection calibration curves with Wavecrest SIA-3000

Figure 3-23 demonstrates that in our test signals PJ and TJ are related by a constant offset. Therefore, we can translate the TJ tolerance testing into PJ tolerance testing once we know the offset. This is important because on ATE we can only accurately generate controllable amount of PJ due to the AWG memory limitation. The final jitter tolerance number we report will be derived from the calibration curve shown in Figure 3-23.

3.2.5 Random Jitter Control

Even though we cannot deliberately inject controllable amount of RJ to the AWG signal due to the size limitation of the AWG memory, we propose an approach to control the RJ in the test signals. The approach utilizes the characteristics of the AWG driver. The offset between the measured DJ and TJ in Figure 3-23 is caused by the intrinsic RJ of the AWG. Because the AWG output signal has a constant rise/fall time, the intrinsic RJ would vary at different output amplitude levels: RJ increases when the signal amplitude decreases and decreases when the signal amplitude increases. Figure 3-24 shows the captured RJ RMS values of 3Gbps test signals at amplitude levels from 230~780mV using the Wavecrest SIA-3000. Therefore, we can control the RJ of our test signals by controlling the amplitude of the AWG output signal, which is programmable on the ATE.

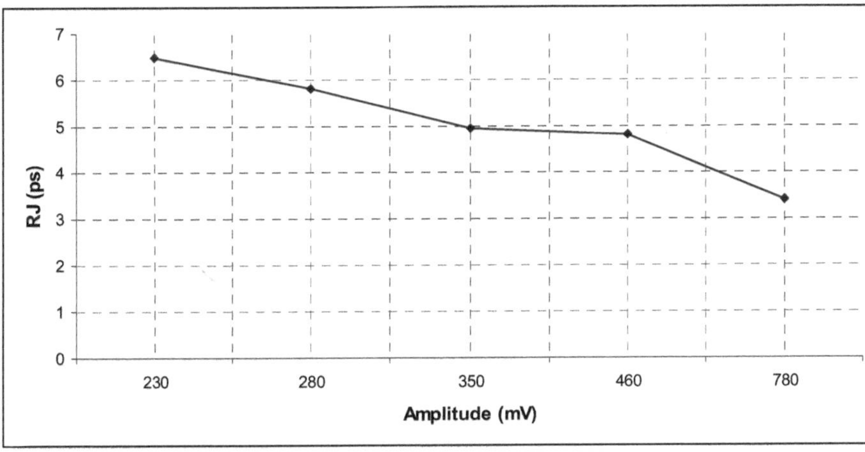

Fig. 3-24. RJ vs. AWG output amplitude

According to the SATA jitter tolerance specification shown in Table 2-4, the difference between the out-of-band DJ and TJ for Gen2 SATA is 0.18UI. The difference suggests that a test signal with a 0.18UI RJ peak-to-peak value is the best candidate to perform both DJ and TJ compliance tests at the same time. This RJ peak-to-peak value translates into a 4.3ps RMS value. According to Figure 3-24, setting the amplitude to around 600mv is the most reasonable setting for the jitter tolerance compliance testing using the generated test signals. This amplitude is also very reasonable as the SATA receiver amplitude range is from 250mv to 700mv.

3.3 Receiver Bit Error Monitoring

When the jittered test signals are applied to the receiver, bit errors may occur in the recovered data. Jitter tolerance testing is done by supplying a test signal with a certain amount of injected jitter to the receiver and then monitoring the bit errors in the recovered data to check whether the BER is below a certain level. The receiver error rate can be monitored using several methods. One approach is to loop back the received parallel data signals to the transmitter and then check the bit errors from the output of the transmitter. This approach usually needs a high speed BERT, which is not available on ATE. The under-sampler on the ATE is not a good candidate for BER measurement. In addition, the transmitter itself might introduce errors, which makes it hard to justify the receiver jitter tolerance test results. Therefore, we need to use alternative approaches. This chapter presents two approaches based on available ATE instruments or DFT features. Chapter 5.2 introduces another approach based on FPGAs.

3.3.1 ATE-based Error Detection

The ATE-based error detection solution brings out the recovered parallel data signals to device pins, and then compares the outputs of these pins with expected values in a digital pattern using digital channels. Figure 3-25 shows the testing configuration of this approach.

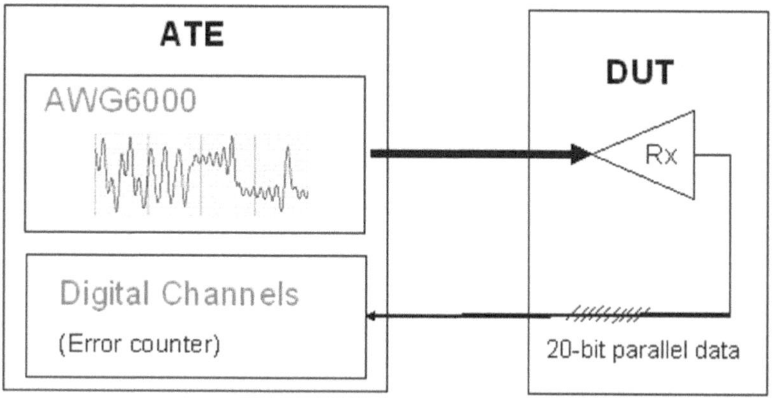

Fig. 3-25. ATE-based BERT

The number of errors on each pin can be accessed by reading back the error counter of the High Speed Digital (HSD) channel associated to the parallel output pin. The error counter is essentially a byproduct of the failure capture memory. It keeps track of how many pattern cycles the fail flag has been asserted from the last HSD reset or counter clear to the time the counter is read. Each failure counter is 16 bits in width. In our devices, the parallel data are 20 bits in width. The maximum number of errors the ATE can track is more than 1 million, which is enough to suppress the statistical variation and allow a generous step size on incrementing the jitter injection amount. This is important because we need to obtain multiple BER points for extrapolation, but the performance (and hence the error rate) can vary from device to device at the same injected jitter level.

One challenging issue of using the parallel data bus for error counting is the synchronization. Because the AWG and digital channels on ATE are in two clock domains, clock synchronization is a priority. Considering the two domains are generated from the same clock source on ATE, clock synchronization can be achieved by properly setting the two clock dividers such that the two frequencies are coherent and the receiver can work correctly. The digital channel strobe is set centered on the parallel data of the receiver output. The byte (i.e. word boundary) alignment is becoming a norm for almost all devices. The frame alignment character is detected by the receiver to ensure that the right sequence of a word (from LSB to MSB) comes out of the parallel bus in a consistent way, from run to run, and from device to device. For pattern alignment, because the delay from the input of the receiver to the output of the receiver varies from device to device, we use a match loop to line up to the repeating PRBS pattern. As shown in Figure 3-26, the match loop skips byte-by-byte to search expected parallel data. Once an expected parallel data sequence is detected (the parallel data bus is aligned to the AWG serial output sequence), the pattern jumps out of the match loop and the HSD channels start checking errors.

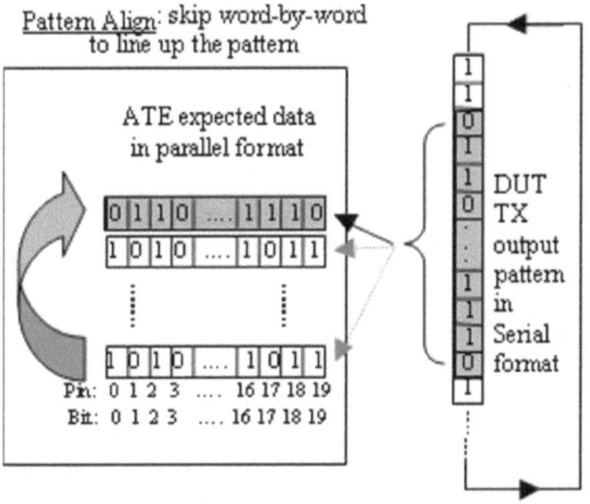

Fig. 3-26. Pattern alignment between the serial data and the parallel data

We can vary the length of the error checking pattern to get the BER with an expected confidence level. A longer length pattern provides a higher confidence level, but takes more test time. By sourcing test signals with different amounts of injected jitter, we can get different bit error rates.

3.3.2 DFT-based Error Detection

The ATE-based BERT is complicated to implement from the test point of view because it involves external synchronization. The BER testing can be simplified by adding DFT features. DFT has been widely used in devices manufactured today in order to reduce testing cost or removing the need for expensive testers. Researchers have proposed Built-in-Self-Test (BIST) techniques specifically tailored for testing common designs, such as bit error checkers, ADCs and PLLs [87], [88], [89].

The idea of DFT-based error detection approach relies on an implementation of a BERT inside the device. The internal BERT includes a pattern generator and an error counter multiplexed with transmitter and receiver latches. Figure 3-27 illustrates the main idea of this approach [87]. The pattern generator generates test sequences, such as the 128-PRBS pattern as we used. When the test signals applied to the receiver have the same sequence as the one generated by the pattern generator, the internal error counter can record the errors that have occurred after the synchronization is achieved.

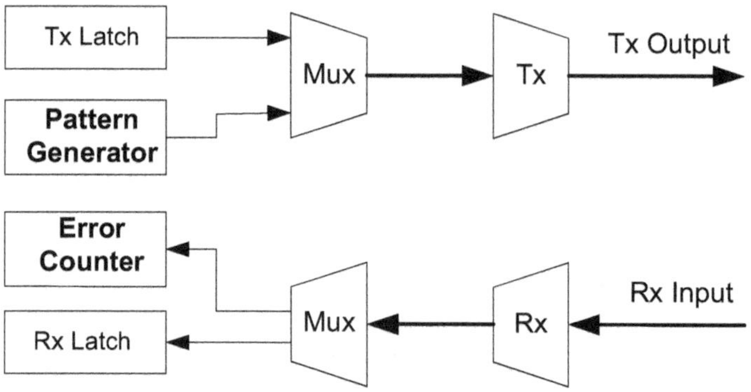

Fig. 3-27. Conceptual illustration of the DFT-based BERT

This approach is widely used in HSSI designs because it simplifies the verification and testing process with little extra design cost. With the built-in BERT, loopback testing can easily be implemented. For receiver function verification, only a small error counter, such as a 4-bit counter, needs to be implemented in the error counter. The receiver function can be verified by just checking the contents of the error counters after the synchronization is achieved: if the number of errors is zero, the receiver functions correctly; otherwise, it fails.

For BER testing, the range of the error counter needs to be big enough to avoid the error counter getting saturated quickly when the error rate is high. The jitter tolerance testing requires us to sweep the input signals with different amounts of injected jitter. The BERs may range from 10^{-5}~10^{-12}. To accommodate such a wide BER range, the error counter needs to be bigger than 10^7. Therefore, the width of the error counter should be 24 bits or more for jitter tolerance characterization. By reading back the values in all the error counter registers, we know the total number of errors. The BER is measured by calculating the ratio between the total number of errors and the total number of tested bits.

3.4 Jitter Tolerance Extrapolation

As discussed in Chapter 3.1.2, we can not use the transmitter bathtub curve fitting technique for receiver jitter tolerance testing. We need to develop a new algorithm for jitter tolerance extrapolation.

3.4.1 Jitter Tolerance Extrapolation Algorithm

The goal of jitter tolerance extrapolation is to predict the jitter tolerance at low BER based on high BER region data. Figure 3-28 illustrates the jitter extrapolation process. We sweep the injected PJ in small increments to get several high BER levels, which can be obtained quickly. In Figure 3-28, the right-bottom plot shows the bathtub curves of these BER levels; the left bottom plot shows the PJ vs BER curve. Because the Q factor and the BER are linked by the inverse error function according to Equation (2-9), we can transfer the measured PJ vs. BER data into PJ vs. Q factor data. According to our jitter tolerance extrapolation algorithm that the relationship between the Q factor and the PJ is linear (discussed later), we can do a Q factor linear fitting based on the PJ vs. Q factor data. The plot in the top left corner is a Q factor fitting result. Based on the fitting result, we can return to predict the PJ tolerance at low BER region through the error function.

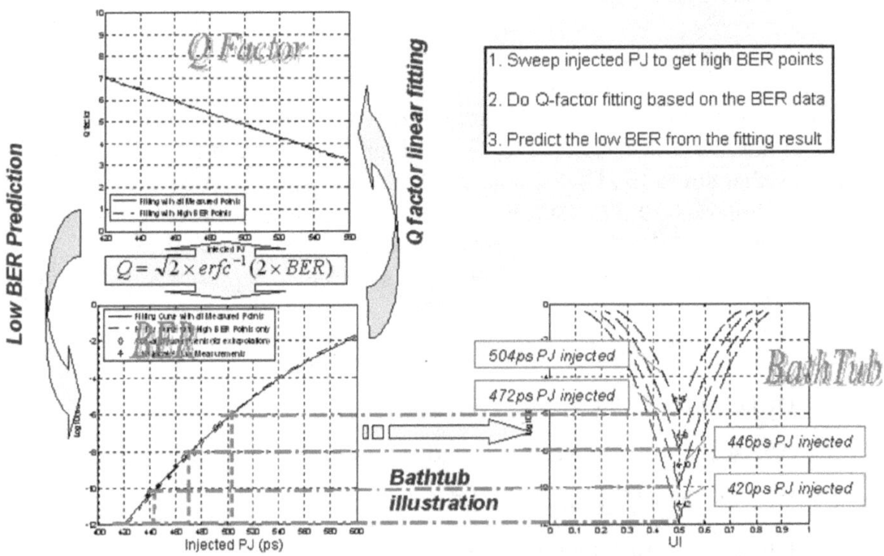

Fig. 3-28. Overview of the jitter tolerance extrapolation via Q-factor fitting

In the above jitter tolerance extrapolation process, every step is straightforward except the Q-factor linear fitting. The fitting is based on the jitter tolerance extrapolation algorithm that there is a linear relationship between the Q-factor and the PJ.

The following reasoning will explain how we derive our jitter tolerance extrapolation algorithm.

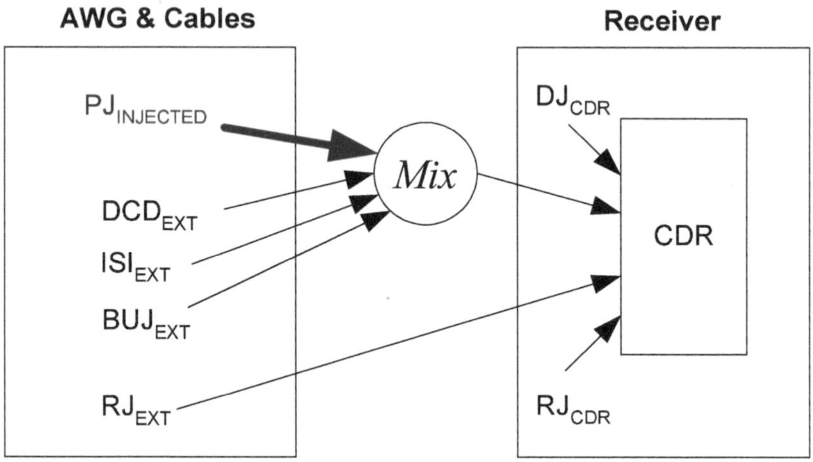

Fig. 3-29. Jitter sources to the CDR

In our testing setup shown in Figure 3-12, we use an AWG to source the test signal and the AWG is connected to the receiver input through RF cables. The jitter sources that stress the CDR come from the AWG, connection cables and the CDR itself. Figure 3-29 illustrates how these different jitter sources act. All the jitter components are convolved together to affect the CDR performance.

We will use in this chapter the following symbols to represent different jitter sources:

- $PJ_{INJECTED}$: the PJ injected in the AWG signals
- DCD_{EXT}: the DCD from the AWG and cables
- ISI_{EXT}: the ISI from the AWG and cables
- BUJ_{EXT}: the BUJ from the AWG and cables
- RJ_{EXT}: the intrinsic RJ of the AWG
- DJ_{CDR}: the intrinsic DJ of the device
- RJ_{CDR}: the intrinsic RJ of the device

Under the "black box" assumption, we have no knowledge about the DJ_{CDR} and RJ_{CDR}. However, we can assume that DJ_{CDR} and RJ_{CDR} are constant for the same device with the same injected jitter frequency because they are determined by the CDR response to the jitter frequency. In [90], a method is proposed to measure the receiver internal jitter that is worth pointing to.

We can also assume that the RJ_{EXT} is a constant when we sweep through different amounts of injected PJ at a constant frequency. This is a reasonable assumption because the RJ in the AWG comes mostly from the sampling clock, which is constant when we change the programmed AWG data samples for PJ injection. The assumption can also be validated by the jitter calibration result shown in Figure 3-23, where RJ_{EXT} is the constant offset between the measured DJ and the measured TJ. Considering RJ_{CDR} and RJ_{EXT} are independent, if we denote the total RJ seen by the CDR with RJ_{TOT}, RJ_{TOT} is constant and we have:

$$RJ_{TOT} = \sqrt{RJ_{EXT} + RJ_{CDR}} \tag{3-1}$$

The DCD_{EXT}, ISI_{EXT} and BUJ_{EXT} are assumed to be constant when the PJ is incremented. This is also a reasonable assumption, because these sources of jitter are mainly determined by the group delay caused by the bandwidth limitation. This assumption is also validated by the constant offset between the PJ and TJ shown in Figure 3-23. Of course, how exactly the PJ combines with the ISI depends on the relative phase relationship, which is unknown inside the DUT. As all the DJ sources are uncorrelated, the total DJ seen by the CDR, DJ_{TOT}, can be expressed as:

$$DJ_{TOT} = (PJ_{INJECTED} + ISI_{EXT} + DCD_{EXT} + BUJ_{EXT}) + DJ_{CDR} \tag{3-2}$$

Under the Q-factor model at the bathtub crossing point (filling up $1UI$) discussed in Chapter 2.2.2, the RJ and DJ components are related to UI by Equation (2-7):

$$UI = DJ + 2Q * RJ_{RMS}$$

where Q is $Q(x)$ defined in Equation (2-1) with $x = BER$, as elaborated earlier in [69].

After substituting Equations (3-1) and (3-2) to Equation (2-7), we obtain the following expression:

$$UI = PJ_{INJECTED} + DJ_{DELTA} + 2Q * RJ_{TOT} \tag{3-3}$$

where DJ_{DELTA} is a constant defined as

$$DJ_{DELTA} = ISI_{EXT} + DCD_{EXT} + BUJ_{EXT} + DJ_{CDR}$$

By rewriting Equation (3-3) to solve for Q and PJ, we obtain the following linear dependencies:

$$Q = C \times PJ_{INJECTED} + S \tag{3-4}$$

$$PJ_{INJECT} = \frac{Q - S}{C} \tag{3-5}$$

where C and S are constants defined by

$$C = -\frac{1}{2RJ_{TOT}} \tag{3-6}$$

$$S = \frac{UI - DJ_{DELTA}}{2RJ_{TOT}} \tag{3-7}$$

Equation (3-4) demonstrates that the Q factor is a linear function of the injected PJ. As discussed in Chapter 2.2.2, the Q factor and BER are related by the complementary error function and shown in Equations (2-8) and (2-9). By substituting Equation (3-4) into Equation (2-8), we have:

$$BER = 0.5 * erfc\ (\frac{C * PJ_{INJECTED} + S}{\sqrt{2}}) \tag{3-8}$$

Therefore, we related BER to a single variable, $PJ_{INJECTED}$ via the classical complementary error function – *erfc*. The curve fitting for this function has matured for decades when applied to transmitter bath tub curve fitting. Even though we cannot use in our case the bathtub curves to represent jitter tolerance extrapolation, the mathematics needed for conducting the jitter tolerance curve fit is still the complementary error function.

Equation (3-8) enables us to estimate BER according to the injected PJ in the test signal. We can extrapolate the PJ tolerance at low BER levels (such as 10^{-12}) once we know the two constant values C and S, which can be obtained using higher BER data (such as 10^{-10} and higher).

3.4.2 Accelerating Jitter Tolerance Characterization

In design validation and device characterization, we need to get the TJ tolerance number at different PVT corners. The test signal calibration results shown in Chapter 3.2.4 enable us to translate the TJ tolerance testing into PJ tolerance testing.

Our scheme is aimed to perform a PJ tolerance BER scan using test signals with different levels of injected PJ. High BER data are collected in the range of 10^{-6} to 10^{-10}. Q-factor values at different BER levels are calculated according to Equation (2-9). Theoretically, we only need two data points to get the two constant values C and S from linear equations in Equation (3-4), according to the equations:

$$C = \frac{Q_1 - Q_2}{PJ_1 - PJ_2}$$

$$S = \frac{Q_2 * PJ_1 - Q_1 * PJ_2}{PJ_1 - PJ_2}$$

However, we still need to observe a large number of errors when testing BER in order to get a high confidence level because of the randomness of the RJ, as explained in Chapter 2.

A better approach, taking into account the confidence level requirements, is to use more data points and perform linear regression fitting over them. In the example to be shown below, jitter tolerance BER scan was performed using 1.5Gbps test signals discussed in Section 3.2.2, and BER data was collected in the range of 10^{-6} to 10^{-11}. The data between 10^{-6} to 5×10^{-9} was considered to be a high BER data and was used to predict the remainder of the points, which are considered as low BER points.

Figure 3-30 shows a linear regression fitting of the Q factor versus injected PJ. Figure (a) is the fitting result based on all measured BER points collected between 10^{-6} and 10^{-11}, while Figure (b) displays the fitting results based on high BER points only.

The difference between the two fitting results is shown in Figure 3-31. The two fitting lines are almost the identical, which demonstrates that the prediction based on high BER points only is very accurate.

Based on the Q factor fitting result, we can now plot the BER curve as a function of the injected PJ, and thus predict the jitter tolerance for a lower BER, e.g. at levels such as 10^{-12}.

Figure 3-32 shows the difference between the BER curve predicted based on the high BER points and the curve fitted with all measured BER points. The discrepancy is found to be very small; from Figure 3-32, we read only 2ps difference at 10^{-12} BER. In this plot, diamond points (high BER data) are used for the prediction; star points are additional real measurements at the lower BER region.

As shown in Figure 3-32, the real measurements in the lower BER range (points labeled with stars) are very close to the predicted curve, which indicates that our BER prediction can successfully match the measurements on real devices under test.

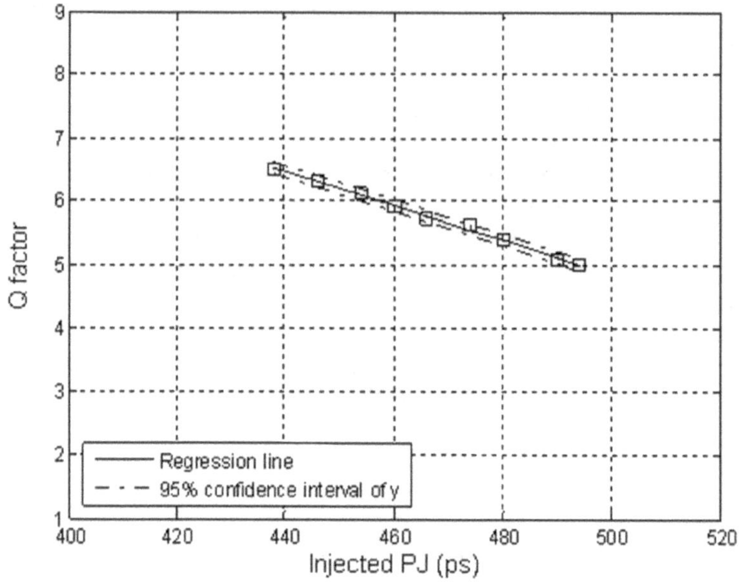

a) Fitting with all BER points

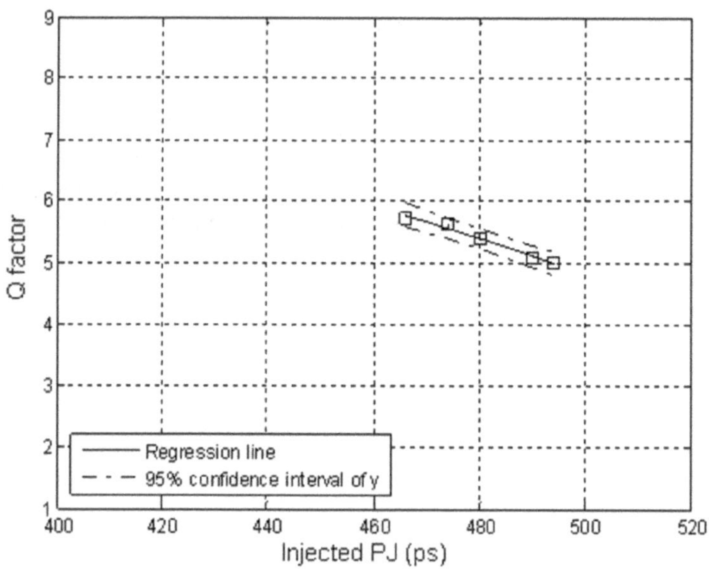

b) Fitting with high BER points

Fig. 3-30. Linear regression of Q factor as a function of the injected PJ

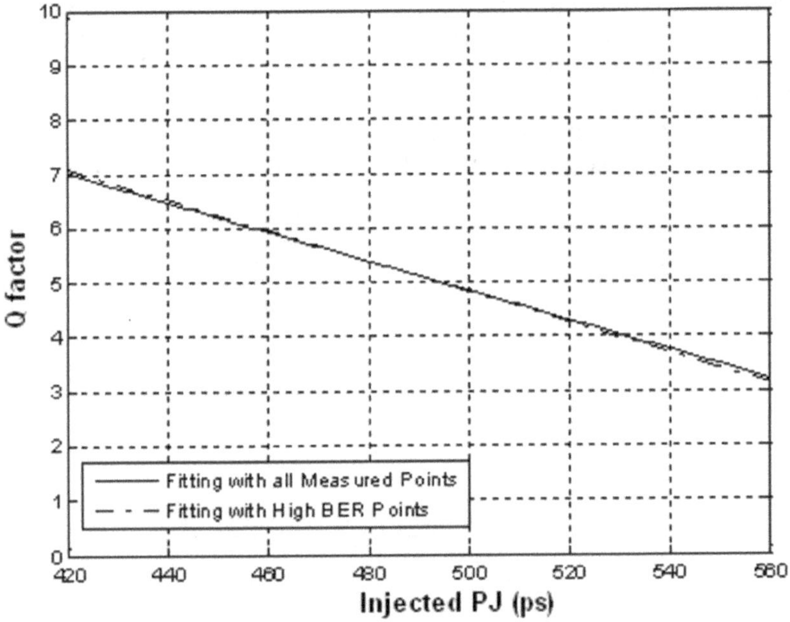

Fig. 3-31. A comparison between the two fitting results

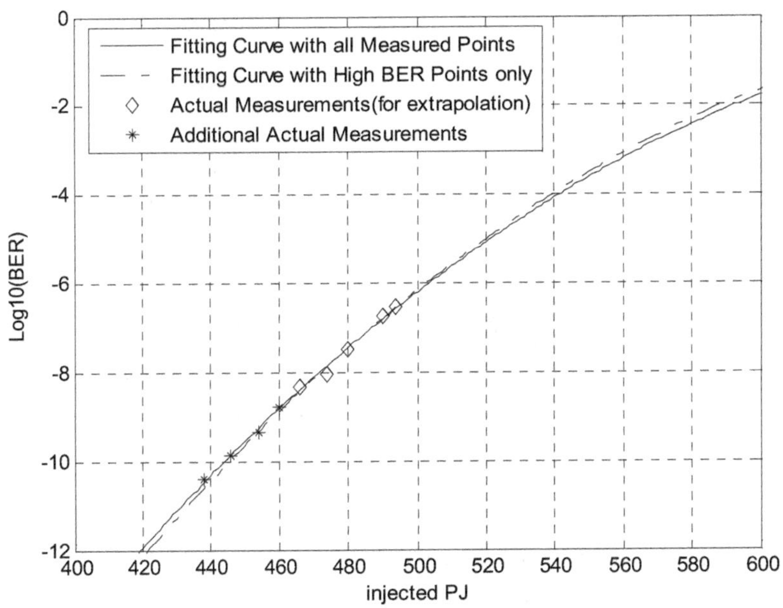

Fig. 3-32. A comparison of BER curve fitting results

The jitter tolerance result presented in Figure 3-32 is in terms of PJ only. To include the DCD, ISI, BUJ and RJ, we need to link this diagram with the jitter injection calibration curves from the Wavecrest SIA-3000.

As shown in Figure 3-23, the delta between the injected PJ and the actual TJ observed by the SIA-3000 is a constant around 92ps. For this particular device under test, the jitter (TJ) tolerance at 10^{-12} BER is 512ps or 0.76UI, while 420ps of PJ is injected in the test signal.

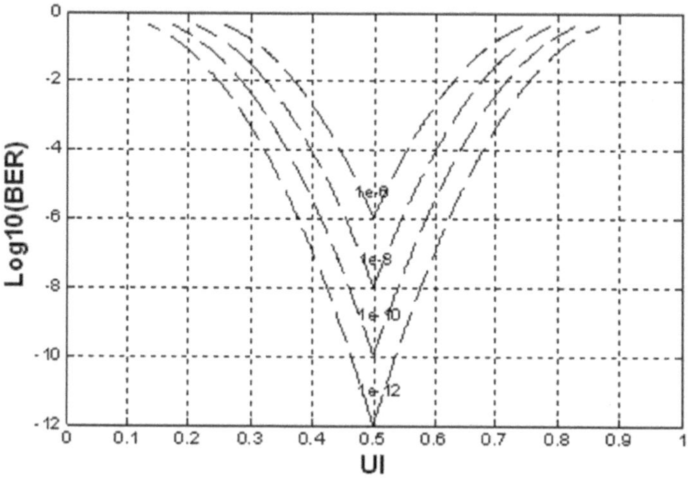

a) Predicted bathtub curves

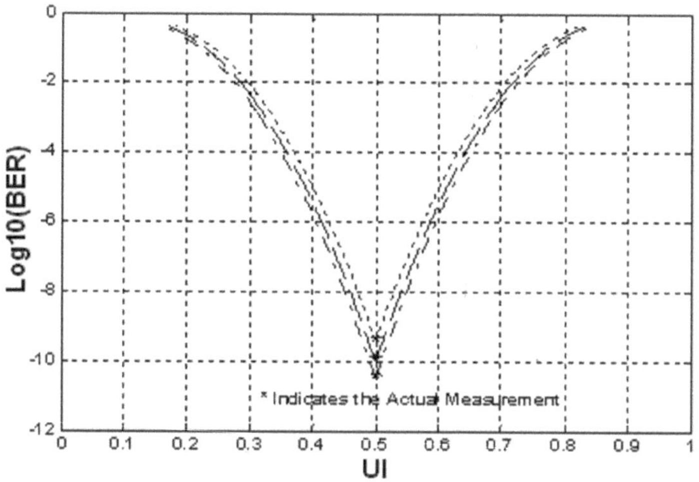

b) Predicted curves vs. actual measurements

Fig. 3-33. Bathtub curve prediction

Figure 3-33 can further help understand the story from the bathtub point of view. Figure (a) conceptually shows that the depth of the bathtub curve moves with different amounts of injected jitter; Figure (b) is a comparison of the predicted bathtub curves to real measurements. This is another way to visualize the accuracy of our model.

We further verified our jitter tolerance extrapolation algorithm at 3Gbps applications and measured the BER down to the 10^{-12} level. For 3Gbps applications, the test signals need to be re-generated and calibrated using the methods discussed in Chapter 3.2. Figure 3-34 shows the calibration results of the generated 3Gbps test signals using a Wavecrest SIA-3000; it plots the measured PJ and TJ values at different injected PJ levels on one tester. As can be seen, from 20ps to 200ps, the measured PJ correlates well to the injected PJ and there is a constant offset between the measured PJ and the measured TJ. In this case, the offset is around 80ps. When the injected PJ is below 20ps, the TJ does not change much due to the noise floor of the AWG. This does not impact us as we will show later that the test signals we need should have PJ values more than 100ps.

Once again, the constant offset between the injected PJ and the measured TJ is caused by the intrinsic RJ of the AWG, and DCD, ISI and BUJ from the AWG and cables. The offset enables us to relate the PJ to TJ, and we can translate the TJ compliance testing into PJ testing, which is needed because we can control the amount of PJ in the test signal.

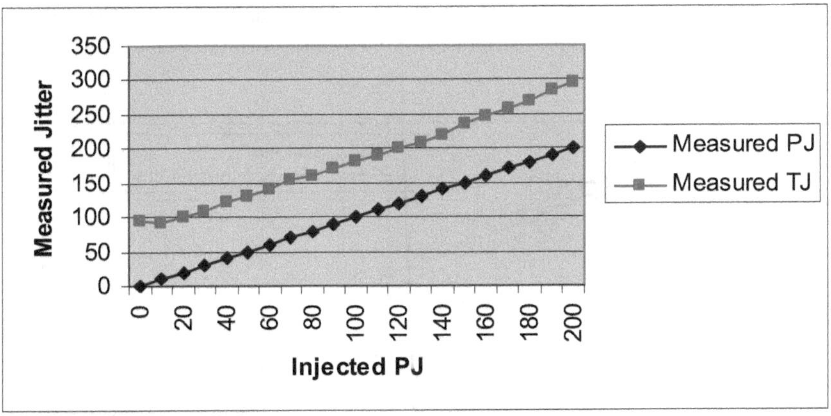

Fig. 3-34. 3Gbps test signal calibration results

Table 3-2 lists an example of measured BER values at different PJ levels using the 3Gbps test signals; the Q-factor values are calculated based on the BER value according to Equation 2-9. When we sweep the injected PJ from 228ps to 216ps, the BER levels decrease from 10^{-6} to 10^{-10}. In this experiment, the bit errors are obtained using the DFT-based error detection approach discussed in Chapter 3.3.2. We use these measured high BER data to extrapolate the BER performance at lower BER levels and then verify the extrapolation results by making BER measurements down to 10^{-12} BER.

Table 3-2. High BER Data for Jitter Tolerance Extrapolation

PJ(ps)	216	218	220	222	224	226	228
BER	2.13E-10	4.37E-10	3.90E-9	2.43E-8	1.05E-7	7.06E-7	2.05E-6
Q	6.24	6.13	5.77	5.46	5.19	4.82	4.60

Figure 3-35 is a linear regression fitting of the Q factor versus PJ based on the measurement results shown in Table 3-2. Based on the Q factor fitting result, we can now plot the BER curve as a function of the injected PJ according to Equation (2-8), and thus predict the jitter tolerance at low BER levels. Figure 3-36 shows the BER curve based on the fitting result from the measurements listed in Table 3-2 (diamonds). It also shows low BER measurements for extrapolation accuracy verification (star points).

As we can see from the data, at 10^{-12} BER there is only 1ps discrepancy between the actually measured PJ tolerance and the extrapolated PJ tolerance based on the BER data above 10^{-10}. We tried the procedure on different devices and the discrepancy is found to be within 2%, while our solution can speed up the characterization over1000 times.

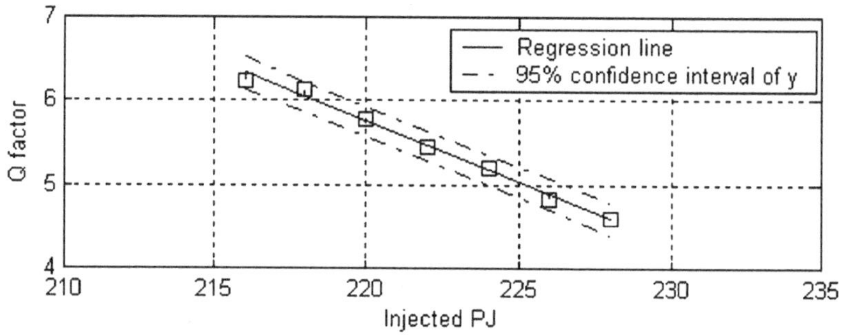

Fig. 3-35. Q factor vs. PJ

The jitter tolerance result presented in Figure 3-36 is in terms of PJ. To translate the PJ tolerance into TJ tolerance, we need to use the jitter injection calibration curves shown in Figure 3-34, where the delta between the injected PJ and the actual TJ observed by the SIA-3000 is a constant around 80ps. For this particular device with BER data listed in Table 3-2, the TJ tolerance at 10^{-12} BER is 292ps or 0.88UI while PJ is 212ps.

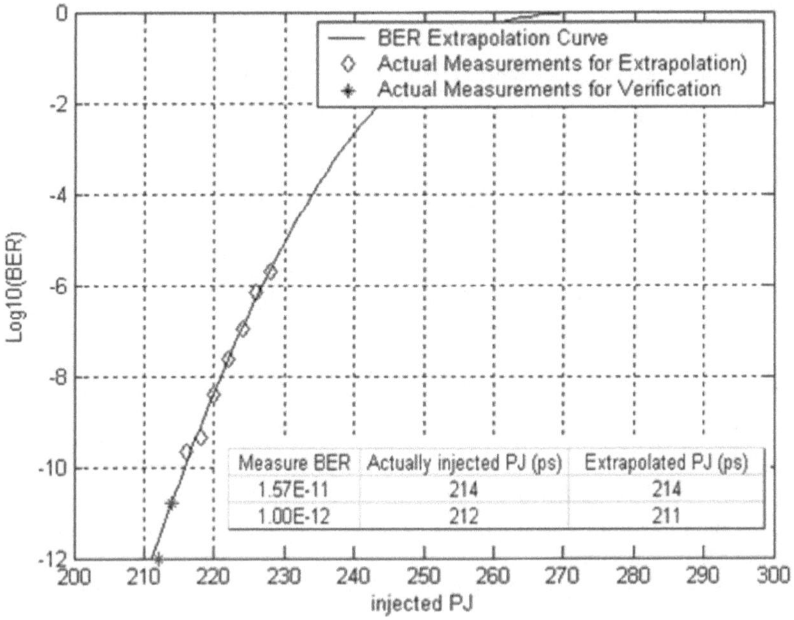

Fig. 3-36. BER extrapolation

3.4.3 Accelerating Jitter Tolerance Compliance Testing

If we directly apply the jitter tolerance characterization technique in production to qualify jitter tolerance compliance, the test time overhead is still a bit high because it involves BER to Q-factor translation, Q-factor fitting and BER extrapolation down to 10^{-12} level. It takes around one second, which is still too long for one parameter testing – on average an SoC device may have hundreds of parameters to test, so the test time of each test is very critical to the final device cost.

Considering that the compliance testing in production is only a go/no-go judgment process, we do not need to know the exact value of the jitter tolerance for each device; we only need to know whether the jitter tolerance of a device is better than the jitter specification defined at 10^{-12} BER.

Instead of extrapolating the jitter tolerance down to 10^{-12} BER and comparing it with the specification, we propose to perform the jitter tolerance compliance test at a higher BER level. For example, we can do the test by qualifying 10^{-6} BER performance. We apply a test signal with a certain amount of injected PJ to the device: if the measured BER of the device is better than 10^{-6}, it passes; otherwise, it fails. For this approach, we need to solve two issues:

- Translating the jitter tolerance specification from 10^{-12} BER level to 10^{-6} BER level
- Translating the TJ specification to a PJ specification

The proposed jitter tolerance extrapolation algorithm can transfer jitter tolerance specifications at different BER levels. Because the Q-factor values at 10^{-12} and 10^{-6} BER levels are known (7.0374 and 4.7534 respectively [38]), according to Equation (3-5), the PJ tolerance difference between 10^{-12} and 10^{-6} BER levels can be calculated by

$$PJ_{10^{-12}-10^{-6}} = \frac{Q(10^{-12}) - Q(10^{-6})}{C} \tag{3-9}$$

which is 25ps in the 3Gbps example in Chapter 3.4.2. According to Equation (3-1) and Equation (3-6), the difference is determined by the RJ in the test signal and the intrinsic RJ of the device that slightly varies from device to device. For each new design, we need to perform the jitter tolerance extrapolation to characterize the PJ difference distribution and use the worst case value (the minimum value) to set test limits for production.

Next, we need to translate the TJ specification into PJ specification because we can only control the amount of PJ in the test signal. The test limit we need to set in production should be based on the amount of the injected PJ. This translation is done according to the offset value between the measured TJ and the injected PJ as shown in Figure 3-23 and Figure 3-34. To do this translation for production, we need to perform the test signal calibration at all the testers because the offset may vary from tester to tester. Figure 3-37 shows the offset at some testers. For these testers, we can claim that the offset between the injected PJ and the actual TJ is at least 70ps.

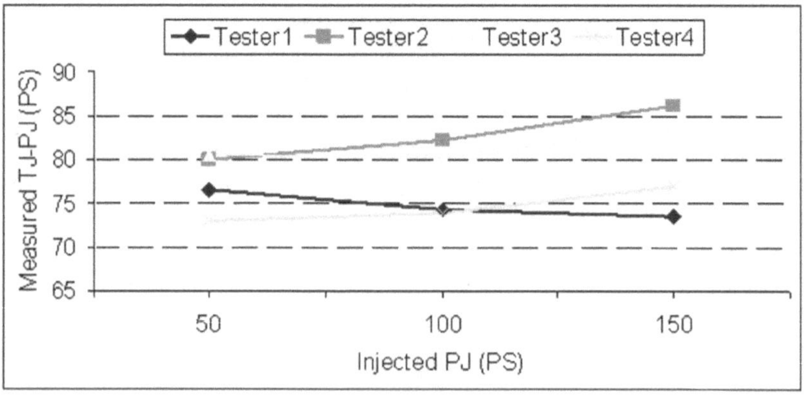

Fig. 3-37. The offset between PJ and TJ at different testers

Based on this offset, we can translate the TJ specification into a PJ tolerance requirement. In the case of SATA II specification, the out-of-band TJ tolerance specification is 200ps at 10^{-12} BER level [22]. We can guarantee the TJ specification compliance by checking the PJ tolerance at 130ps: if a device can tolerate 130ps PJ at 10^{-12} BER level, we can guarantee that the device meets the SATA jitter tolerance specification. Even though this might slightly overstress devices on some testers (such as Tester2 and Tester3), this is acceptable as long as it does not cause yield issues.

According to jitter translation Equation 3-9, the PJ difference between 10^{-12} and 10^{-6} BER levels is 25ps. Because the PJ tolerance requirement at 10^{-12} BER level is 130ps, the PJ tolerance limit should be set to 155ps at 10^{-6} BER level. We can source a test signal with 155ps injected PJ to the receiver and check 10^{7} bits of recovered data. If no errors are detected, this device is classified as a good one; otherwise, it fails the jitter tolerance compliance test.

3.4.4 Discussion

In the proposed acceleration scheme for jitter tolerance qualification, the injected jitter calibration and the jitter tolerance extrapolation are the two key techniques that we employ. When applying the scheme to production testing, we need to especially pay attention to them.

Please note that we need to calibrate the test signals on all testers to ensure that the difference between the injected PJ and the measured TJ is bigger than the offset that we used to derive the test limit, which is set to 70ps in the 3Gbps example of SATA devices. If the offset is smaller than this, we need to tighten our test limit accordingly.

At the same time, we also need to keep a close eye on the possible yield loss because we can overstress devices on some testers, such as Tester2 in Figure 3-37, where the jitter in the test signal is 10ps more than the injected jitter that we need to use to stress the device. This should not cause issues because the design margin normally is big enough to accommodate it.

Another source that provides extra margin for the test is the fact that in practical testing we usually classify devices with errors between 1 and 10 out of 10^{7} bits as bad devices. Actually, they might be classified as good ones as they meet 10^{-6} BER performance, but with limited confidence level. Because of the extra margin, we have a high confidence level that the good devices meet the jitter tolerance requirement.

In addition, we need to do the jitter specification translation (from 10^{-12} to 10^{-6} BER levels) based on devices that can cover the products to be tested, such as devices from all process corners. Doing this from one device may not be enough. The good thing is that we only need to do this once for every new design.

Even though the experiment is conducted primarily on Teradyne AWG6000, the jitter tolerance extrapolation technique is generic and can be used on any platform that has jitter injection capability and that can perform BER testing. The technique can rapidly report the actual jitter tolerance value for characterization, or qualify a jitter tolerance specification in time that makes it practical to test massively manufactured devices.

3.5 Other Applications of the New Method

3.5.1 Jitter Transfer Characterization

In previous experiments, we demonstrate injecting PJ at a single frequency to investigate the receiver jitter tolerance. We can extend our experiments to investigate the jitter transfer characteristics of the PLL by applying test signals with different jitter frequencies.

As we discussed in Chapter 3.1.1, the jitter tolerance performance of a CDR is mainly determined by the PLL. The PLL has low-pass characteristics, as shown in Figure 3-3: the recovered clock can track the in band jitter in the input data, but cannot track the out-of-band jitter. We can employ the fact that we can routinely perform the PLL jitter transfer function characterization on an ATE using the test signals that we generated. Figure 3-38 shows the test setup for such characterization tests.

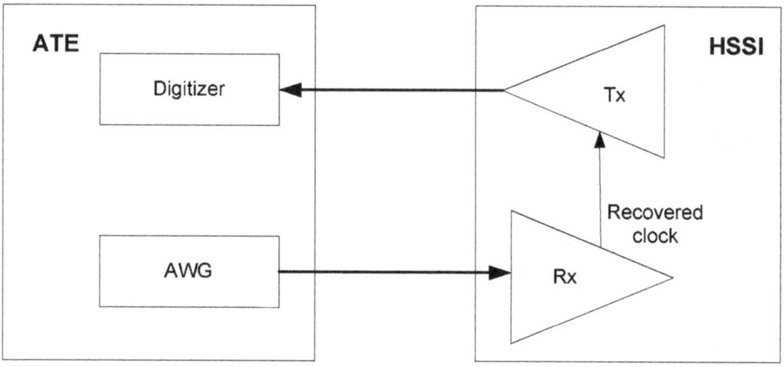

Fig. 3-38. Test setup for the PLL jitter transfer characterization

In the test setup, the AWG sources test signals to the receiver input. The test signals from the AWG all have a known constant amount of the injected PJ jitter. The clock recovered at the receiver side is then used by the *transmitter* to align its own output. A digitizer on the ATE is finally employed to capture the PJ in the transmitter output. In this way, the resulting jitter transfer function is readily observed from the sent data stream.

The PLL jitter transfer function can be derived by the following scheme. One needs to undertake the sweeping of the injected PJ frequencies across the frequency range. Then, the relation between the two jitters is simply obtained by comparing the injected PJ in the AWG with the captured PJ from the transmitter output.

Figure 3-39 shows an example of the PLL jitter transfer experimental result, where the normalized transmitter output jitter is the ratio of the transmitter output PJ to the receiver input PJ.

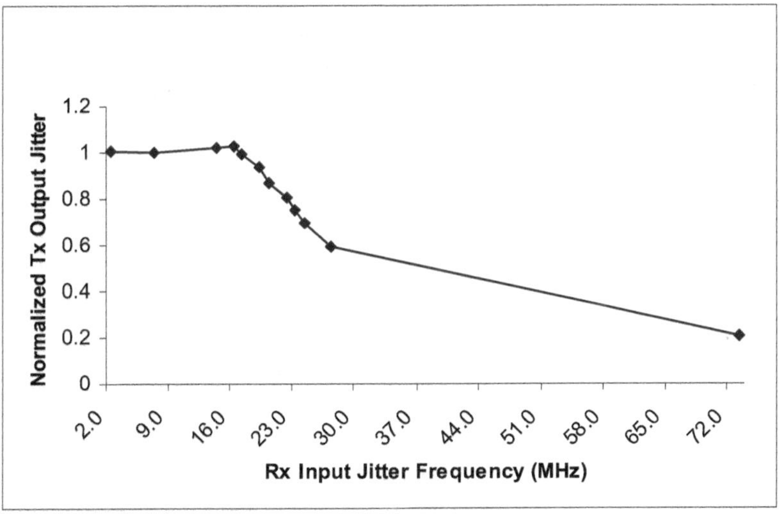

Fig. 3-39. Measured PLL jitter transfer characteristics

As we can see from the plot, when the frequency of the receiver input jitter is bellow the receiver PLL bandwidth, which is around 20MHz in the above example, the transmitter output jitter simply tracks the receiver input jitter. This is true because the clock recovered at the receiver side faithfully tracks the jitter in the AWG signals and therefore passes the injected jitter directly to the transmitter output.

On the other hand, when the jitter frequencies are above the PLL bandwidth, the clock recovered at the receiver cannot track the injected jitter. Therefore, the out-of-band jitter gets naturally attenuated at the transmitter output, with the exact relations depending on the PLL type and its bandwidth.

3.5.2 CDR Characteristics Analysis

Another application of our method is in the exploration of the jitter tolerance frequency characteristics of the CDR. We generate test signals with injected jitter at different jitter frequencies and then get the jitter tolerance performance of the receiver at different jitter frequencies.

Figure 3-40 shows an example of the frequency response of a CDR that we obtained. As we can observe, the frequency response is not flat. It has higher jitter tolerance at low frequencies. When the frequency increases, the jitter tolerance decreases and reaches to a minimum around 50MHz, for a given SATA application.

This type of frequency sensitivity investigation is extremely time-consuming with traditional jitter tolerance test methods. It takes days to accurately characterize a device using traditional techniques. Our proposed accelerated jitter tolerance test scheme can significantly reduce the time needed for this type of characterization.

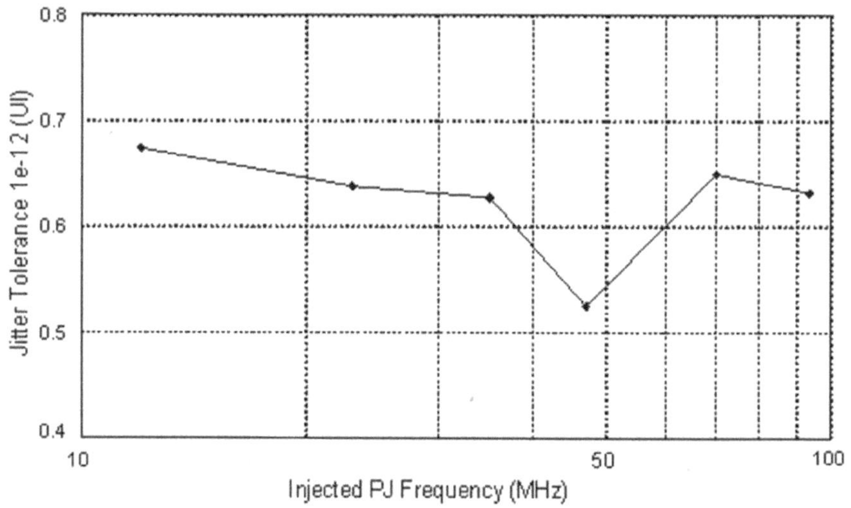

Fig. 3-40. Jitter tolerance frequency response

In addition, our proposed scheme provides a method to develop CDR models. From the testing point of view, the CDR is just a black box. With our scheme, we can easily stress the CDR using test signals with different jitter frequencies and jitter profiles and check its response.

An important conclusion is that, based on this kind of experiments, it is possible to make an accurate CDR model through which the performance of the CDR can be predicted. This is very helpful for the designer to choose the right CDR and for the test engineer to set the proper test condition in production. Without accelerating the receiver jitter tolerance characterization, this achievement would not be possible for massively manufactured and marketed devices, which are the hallmark of the modern integrated circuits and systems.

4 Transmitter Jitter Extractions on ATE

Abstract *In this chapter, we introduce the details of how the transmitter jitter can be decomposed in a way that is fast and accurate enough to be implemented in a production environment. The jitter can be extracted from either the time domain or the frequency domain. We also propose a hybrid approach to improve accuracy while keep test time short. The advantages and disadvantages of each of the proposed method are discussed, such that the best method can be selected for task at hand. Finally, a set of experiments is undertaken to validate the proposed approaches.*

4.1 Introduction

Transmitter jitter testing has been investigated for years. There are quite a few bench instruments that can make the jitter measurements accurately [91], [92]. There are also some ATE platforms that have transmitter jitter testing capability. However, these solutions have limitations: either their throughput is too low or their accuracy needs to be improved. In addition, almost every existing solution has its own proprietary algorithms that are either protected by patents [93] against wider use or unknown to the public.

To overcome these limitations, we research the transmitter jitter testing on ATE and present a systematic solution for multiple Gbps transmitter jitter characterization and production testing. Our under-sampling based approach can extract jitter either from edge histograms in time domain or from the jitter spectrum in frequency domain. The two approaches can also be combined to achieve more accurate test results.

Y. Fan, Z. Zilic, *Accelerating Test, Validation and Debug of High Speed Serial Interfaces*,
DOI 10.1007/978-90-481-9398-1_4,

4.1.1 Transmitter Jitter Testing Overview

As discussed in Chapter 2.2.1, most HSSI standards, such as SATA and Fiber Channel, define DJ and TJ specifications separately. Table 4-1 lists the transmitter out-of-band jitter specifications for the SATA II [22]. The TJ of the SATA transmitter should not exceed 0.37UI at 10^{-12} BER level and DJ should not exceed 0.19UI. Though the SATA specification does not specify the RJ limit, we can get the limit by assuming that all TJ is contributed by RJ. The RJ with 0.37UI peak-to-peak value at 10^{-12} BER level translates into a RJ RMS value of 0.026UI, or 8.8ps at 3Gbps data rate and 4.4ps at 6Gbps data rate. The RJ RMS value is usually used to estimate TJ at 10^{-12} BER level, as it is impossible to directly measure TJ at this BER level in volume production due to long test time – it takes tens of minutes to take a single measurement even at 6Gbps data rate.

Table 4-1. Transmitter Jitter Specifications for SATA Gen2

TJ	DJ	RJ (RMS)*
0.37UI	0.19UI	0.026UI or 8.8ps@3G, 4.4ps at 6G

*Deduced from the TJ specification

To economically apply the above test limits in production, we need the jitter test to have the capabilities of:

- Separating jitter components
- Achieving accuracy in sub-picoseconds
- Having the test done in milliseconds

Unfortunately, there is currently no solution on ATE that meets these criteria, even though the jitter measurement and decomposition have been investigated for years [94], [95]. Popular jitter testing solutions include Bit Error Rate Testers (BERT), histogram-based Oscilloscopes, and Time Interval Analyzers (TIA) [34]. These solutions are commonly used for design validation and characterization on bench. However, we cannot directly apply them for at-speed testing in production because of the low throughput.

There are not many systems right now that can do multi-gigabit devices jitter compliance testing in production. Many jitter test solutions are based on extra on-chip circuitry, or add-on modules [40], [41], [96], [97], [98], [99]. The applications of these solutions are limited either by their low throughput, low accuracy, or high design complexity of the device or of the loadboard. Because of these limitations, pure ATE-based solutions are preferred in production because of their high portability and high throughput.

One approach to provide more and higher-end test functionality is to utilize multi-gigabit signal generators and digitizers. The generators and digitizers are becoming available as fully-integrated ATE instruments. One example is the GigaDig on Catalyst/Tiger ATE from Teradyne [84]. The GigaDig is a digitizer, capable of capturing analog signals with a time resolution better than 1ps. With this kind of instruments, it has become feasible to perform multi-gigabit devices jitter test on ATE [4], [86], even though systematic jitter extraction algorithms on ATE have not matured.

In [100], a transmitter jitter test solution is proposed based on the high-speed digital pins of the Agilent's 93000 ATE. By shifting the compare strobes in the timing axis and level threshold axis, this approach first builds a bathtub curve, and then applies the jitter separation algorithm.

However, the test economy of this solution needs to be improved: it takes close to 1 second to perform the test even with 2ps resolution. Please note that as jitter is just one of hundreds parameters to be tested on an average device, and the testing need to be performed for millions of potentially faulty devices, one second spent for one such test procedure is still too long to be practical in the ATE environment. In addition, the test accuracy also needs to be improved: RJ was here around 1ps higher than the bench result.

In [86], an SATA test solution on ATE is presented, which includes transmitter jitter testing. However, the transmitter jitter testing scheme in [86] is not very accurate; it reports higher RJ (1~2 ps) and also the higher TJ (20ps) than the bench equipment does. Finally, one can further observe that the test parameters in this solution can still be further optimized to achieve better test economy, as the cost incurred during the testing phase is a significant part of the overall cost of the device.

4.1.2 Proposed Solution

In this chapter, we present a new transmitter jitter testing solution based on a high-bandwidth digitizer on ATE [4]. We sensibly make the test setup and develop jitter extraction procedures suitable for running the test fast and accurately. With the current ATE instrument, we achieve sub-picosecond jitter accuracy and can finish the whole transmitter testing in 100ms, which no one else has ever achieved in an ATE environment to our best knowledge. The whole solution has been verified at data rates up to 6Gbps applications.

Better performance and higher data rate applications are attainable using our solution with the advances in ATE instruments in the future – they are only limited by the bandwidth and timing resolution of the digitizer. The accuracy of the proposed solution was verified by both bench equipment and by the ATE itself. More critically, the proposed test procedures have already been applied in volume production.

In the remainder of this chapter, we first describe the principles of setting instruments and test parameters for data acquisition. Then we present the details of the data processing – how jitter is extracted and decomposed in both the time domain and the frequency domain, as well as how the two ways can be combined to provide more accurate information on the jitter sources. After that, the experimental results are presented and the advantages and limitations of each of the methods are discussed.

4.2. Test Setup for Data Acquisition

The digitizer we use for the transmitter testing is GigaDig. The GigaDig is a fully integrated ATE digitizing instrument with a typical under-sampling bandwidth over 9 GHz [84]. Its input voltage range is 64mv to 1.024v, capable of covering most HSSI standards. This chapter concentrates on the SATA transmitter, which has an output range of 400mv~700mv.

With a 1 Mega sample memory and with 12-bit digitizing resolution, the GigaDig can perform all transmitter function and parameter tests with a single capture of the transmitter output. The following discusses how we sensibly set up the GigaDig for data acquisition.

4.2.1 Overview of the Test Setup

The test setup is shown in Figure 4-1. The ATE provides a reference clock signal *tx_ref_clk* to the transmitter; a PLL in the transmitter then locks the transmitter output rate to the reference clock. In our applications, the ideal *tx_ref_clk* is around 30MHz and the transmitter output data rate F_{data} can be 1.5G, 3G or 6Gbps. The GigaDig captures the transmitter output with an under-sampling rate *Fs* between 5 Mega Samples per second (MS/s) to 10 MS/s.

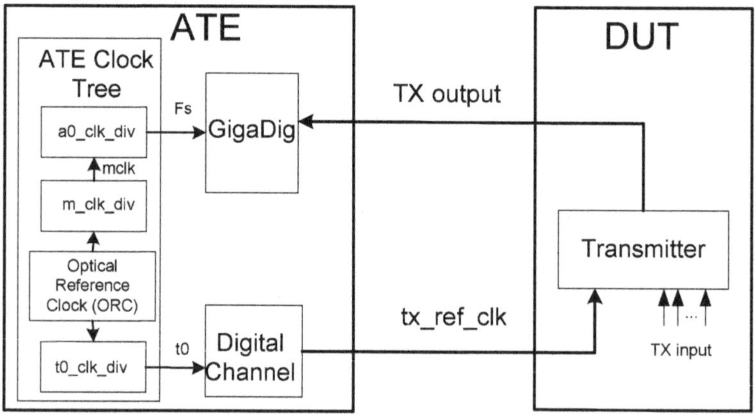

Fig. 4-1. Transmitter test setup for data acquisition

The under-sampling technique has been already used in high-speed testing for years [94]. This technique first captures the output signal of the DUT at a sampling rate Fs that is lower than the output data rate F_{data}, and then shuffles the captured samples in a predetermined manner.

The shuffled output is a sequence of samples that would have resulted from sampling the data at a much higher frequency. The effective sampling rate F_{eff} is defined by

$$F_{eff} = F_{data} * N_{SPB} \qquad (4\text{-}1)$$

where N_{SPB} is the Number of Samples Per Bit (SPB) in the shuffled data.

4.2.2 Principles of Clock Settings

Although the under-sampling principle is simple, the challenge for transmitter jitter testing is to properly set test parameters for data acquisition and to extract jitter information from captured samples.

To capture within reasonable test time the transmitter output waveform with an adequate resolution for jitter decomposition, we need to properly set the following parameters:

o Test pattern length $L_{pattern}$

o Effective sampling rate F_{eff}

o Undersampling rate F_s

In order to set the clock properly to capture the transmitter output in an undersampling environment, we need to first determine the test pattern. To provide adequate test coverage, the test pattern length $L_{pattern}$ should be at least 20 bits because the width of the parallel data to the transmitter input is 20.

On the other hand, the test pattern length should be kept as short as possible in order to save test time and also to simplify the subsequent data processing. For these reasons, we choose a 20-bit test pattern 00000111110101010011. Note that this pattern includes both high density and low density transitions, with a total of eight edges. The transmitter output (GigaDig input) fundamental frequency F_{DUT} is defined by

$$F_{DUT} = \frac{F_{data}}{L_{pattern}} \qquad (4\text{-}2)$$

which generates a 150MHz F_{DUT} when the data rate F_{data} is 3GHz and the pattern length is 20.

Our experimental data in Section 4.4.3 of this chapter also demonstrates that the 20-bit test pattern is a very good choice for transmitter jitter testing. It generates similar RJ and DJ compared to a 128-bit PRBS pattern and a clock pattern. The 20-bit pattern provides reasonably good coverage in a production environment.

The required effective sampling rate F_{eff} is determined by the target test accuracy. To achieve a jitter measurement resolution better than 1ps, we need to have the effective sampling resolution better than 1ps, which corresponds to an effective sampling rate higher than 1000GHz. For 3Gbps data signals, this translates into capturing at least 333 samples per data bit. To leave some margin and also keep the test time short, we choose to capture 400 samples per bit, which gives

$$N_{SPB}=400$$

$$F_{eff}=1200G$$

The under-sampling rate Fs needs to be calculated based on the GigaDig input fundamental frequency F_{DUT} and the effective sampling rate F_{eff}. In order to capture samples coherent with the input signal, we need to satisfy the equation

$$\frac{1}{F_s} = K\frac{1}{F_{DUT}} + \frac{1}{F_{eff}} \tag{4-3}$$

where K is the number of cycles of F_{DUT} slipped before the next sample. According to Equations (4-1) and (4-2), Equation (4-3) can be expressed as

$$\frac{1}{F_s} = \frac{1}{F_{data}}(K * L_{pattern} + \frac{1}{N_{SPB}}) \tag{4-4}$$

As shown in Figure 4-1, ATE provides to the transmitter a reference clock tx_ref_clk with a frequency $F_{tx\_ref\_clk}$. The PLL in the transmitter sets the transmitter output data rate by multiplying the reference clock by an integer M_{d2ref}. Therefore, we have

$$F_{data} = F_{tx\_ref\_clk} * M_{d2ref} \tag{4-5}$$

where M_{d2ref} equals 100 for 3Gbps applications and 200 for 6Gbps applications while $F_{tx\_ref\_clk}$ is around 30MHz. For example, if the reference clock is set to be exactly 30MHz, the output data rate of the transmitter F_{data} would be exactly 3GHz.

However, the ATE cannot source a clock signal exactly at 30MHz because this clock is derived from the Optical Reference Clock (ORC) divided by $t0\_clk\_div$, where ORC is around 50,000THz and $t0\_clk\_div$ can only be an integer. Similarly, Fs is also derived by dividing the ORC. According to Figure 4-1, we have

$$F_{tx\_ref\_clk} = \frac{ORC}{t0\_clk\_div} \tag{4-6}$$

$$F_s = \frac{ORC}{m\_clk\_div * a0\_clk\_div} \tag{4-7}$$

Based on Equations (4-5), (4-6), and (4-7), we can rewrite Equation (4-4) in the following way:

$$\frac{m\_clk\_div * a0\_clk\_div}{ORC} = \frac{t0\_clk\_div}{ORC * M_{d2ref}}(K * L_{pattern} + \frac{1}{N_{SPB}}) \tag{4-8}$$

By re-organizing Equation (4-8), we can get

$$\frac{t0\_clk\_div}{m\_clk\_div} = \frac{a0\_clk\_div * M_{d2ref} * N_{SPB}}{K * L_{pattern} * N_{SPB} + 1} \tag{4-9}$$

Equation (4-9) is the one that we use to determine the values of all the clock dividers (*t0_clk_div*, *a0_clk_div* and *m_clk_div*) in order to capture the transmitter output with the expected resolution. Equation (4-9) applies to all data rates and test patterns. However, if the data rate and/or reference clock is different, M_{d2ref} needs to be adjusted accordingly. In addition, $L_{pattern}$ is determined by the length of the test pattern.

4.2.3 Test Setting Parameter Calculations

Equation (4-9) provides the principle of setting the clock dividers. As an example, we first demonstrate how the clock dividers are set in a 3Gbps application using the 20-bit test pattern. The test setup of the application is shown in Figure 4-1. The reference clock *rx_ref_clk* needs to be around 30MHz; the intended effective sampling rate F_{eff} is 1200GHz; the ATE requires that *mclk* needs to be between 160MHz to 200MHz and the under-sampling clock *Fs* needs to be between 5MHz and 10MHz [84].

If we want *Fs* to be around 7.5MHz, one choice is to set *a0_clk_div* to be 26 and *mclk* to be around 195MHz. By replacing M_{d2ref} with 100, N_{SPB} with 400 and $L_{pattern}$ with 20 for the 3Gbps application, we can then simplify Equation (4-9) to equal the following ratio

$$\frac{t0\_clk\_div}{m\_clk\_div} = \frac{1040000}{8000 * K + 1} \tag{4-10}$$

Basically, Equation (4-10) is a transformation of Equation (4-3) with pre-set parameters based on the application.

The starting observation is hence that we can guarantee the coherent sampling by satisfying Equation (4-10). Our next goal is to get the *K* value in Equation (4-10). Considering that the *tx_ref_clk* should be around 30MHz and that *mclk* is expected to be around 195MHz, we can obtain the rough estimates of the *t0_clk_div* and *m_clk_div*:

$$t0\_clk\_div\_rough = \frac{ORC}{30MHz} \approx 1666666666$$

$$m\_clk\_div\_rough = \frac{ORC}{195MHz} \approx 256410256$$

Therefore, we have

$$\frac{t0\_clk\_div}{m\_clk\_div} \approx \frac{t0\_clk\_div\_rough}{m\_clk\_div\_rough} \approx 6.5 \qquad (4\text{-}11)$$

According to Equations (4-11) and (4-10), we can solve K in Equation (4-10) and K=20 gives the best approximation for the expected *mclk* and *tx_ref_clk* frequencies. When K=20, Equation (4-10) becomes

$$\frac{t0\_clk\_div}{m\_clk\_div} = \frac{1040000}{160001} \qquad (4\text{-}12)$$

Because *t0_clk_div* and *m_clk_div* need to be integers and *t0_clk_div* should be around 166666666, we can choose

$$t0\_clk\_div = 1040000 * 1602 = 1666080000 \qquad (4\text{-}13)$$

$$m\_clk\_div = 160001 * 1602 = 256321602 \qquad (4\text{-}14)$$

where 1602 is the floor of the ratio between *t0_clk_div_rough* and 1040000 (1666666666/1040000 $\approx$ 1602.564).

Using the clock divider values shown in Equations (4-13) and (4-14) and the *a0_clk_div* value (pre-set to 26), we can generate the clocks that enable us to capture the 3Gbps signal with 400 samples per bit.

We now illuminate how this line of reasoning is used in practice. For this purpose we show examples of practically obtainable parameters for realistic clock rates of modern devices. First, Table 4-2 presents a concise summary of the parameters used for the 3Gbps data capture. Then, Table 4-3 summarizes the actual under-sampling frequency and the transmitter reference clock frequencies used in this case.

It is important to keep in mind that the above derived clock divider values present just one combination of the possible settings that can be used to capture waveforms with the expected resolution. Based on the same procedure, we can equally obtain other clock divider settings if we pre-set *Fs* and *mclk* to different frequencies, only under the condition that the derived clocks are held within their valid ranges.

Table 4-2. Parameter Settings for 20-bit Pattern 3Gbps Data Capture

Parameter	Description	Value
F_{data}	Transmitter output data rate	3GHz
$L_{pattern}$	Length of the data pattern	20
M_{d2ref}	Transmitter PLL multiplier	100 (30MHz ref clock)
F_{DUT}	Fundamental frequency of Tx output	150MHz
N_{SPB}	Number of samples per bit	400
F_{eff}	Effective sampling rate	1200GHz
a0_clk_div	a0 clock divider	26
t0_clk_div	t0 clock divider	1666080000
m_clk_div	mclk divider	256321602

Table 4-3. Actual Clock Frequencies for the 3Gbps Data Application

Parameter	Description	Value
Fs	Undersampling clock frequency	≈7.502594MHz
$F_{tx\_ref\_clk}$	Tx reference clock frequency	≈30.010564MHz
F_{mclk}	mclk frequency	≈195.067445MHz

Based on the same principles applied to set the parameters in the explicit procedure demonstrated in the 3Gbps example, we can obtain the proper parameters for other applications as well. Table 4-4 enumerates the parameters for 6Gbps applications using a 20-bit test pattern. Further, Table 4-5 lists the parameters for 5.5Gbps applications, where a 20-bit test pattern and a 27.5MHz reference clock are used.

Following are few highlights of parameters that need to be changed for some other applications:

- If F_{data}=5.5GHz and the test pattern is 20 bits in length, then F_{DUT} is 275MHz
- If F_{data}=3GHz and the test pattern is 128 bits in length, F_{DUT} is 23.4375MHz
- If the expected reference clock is different, *t0_clk_div_rough* is then different

Table 4-4. Parameter Settings for 6Gbps Data Capture

Parameter	Description	Value
F_{data}	Tx output data rate	6GHz
$L_{pattern}$	Length of the data pattern	20
M_{d2ref}	Tx PLL multiplier	200 (30MHz ref clock)
F_{DUT}	Fundamental frequency of Tx output	300MHz
N_{SPB}	Number of samples per bit	400
F_{eff}	Effective sampling rate	2400GHz
a0_clk_div	a0 clock divider	26
t0_clk_div	t0 clock divider	1666080000
m_clk_div	mclk divider	256320801

Table 4-5. Parameter Settings for 5.5Gbps Data Capture

Parameter	Description	Value
F_{data}	Tx output data rate	5.5GHz
$L_{pattern}$	Length of the data pattern	20
M_{d2ref}	Tx PLL multiplier	200 (27.5MHz ref clock)
F_{DUT}	Fundamental frequency of Tx output	275MHz
N_{SPB}	Number of samples per bit	400
F_{eff}	Effective sampling rate	2200GHz
A0_clk_div	a0 clock divider	26
T0_clk_div	t0 clock divider	1817920000
M_clk_div	mclk divider	279680874

The last test setup parameter we need to choose is the required number of samples N_{total}. The required number of samples can be derived from the pattern length and the effective sampling rate. To build edge transition histograms and to acquire their statistical properties for jitter extraction, we need to capture at least a certain number of cycles of the test pattern. For a statistics process, the Standard Error (SE) of the mean SE_m is defined by

$$SE_m = \frac{\delta}{\sqrt{n}} \tag{4-15}$$

where δ is the standard deviation of the population and *n* is the size of the samples [101] [102].

From the statistical confidence point of view, the larger the sample size is, the less likely the error is. On the other hand, we need to minimize the number of samples in an ATE environment in order to minimize the test cost. To balance well the tradeoffs in this case, we decided to capture 20 cycles of the 20-bit test pattern.

As we will discuss later, we use all the transition edges to calculate the RJ. There is the total of 160 edges in 20 cycles of the 20-bit test pattern. Therefore, the statistical error of the RJ is less than 8% according to Equation (4-15). This is an acceptable error bound for our test in light of the existing margins throughout the high-speed serial interface test and characterization.

Another factor that we need to consider when determining the total number of samples is the Fast Fourier Transformation (FFT) requirement. We will need FFT later for jitter decomposition in the frequency domain. Twenty cycles of the 20-bit pattern with 400 samples per bit translate into 160k samples in total. The derived number of samples is hence seen to satisfy the FFT requirement.

4.3. Jitter Extraction

Jitter is extracted from the transmitter output waveform captured with the test setup and the parameters discussed in Chapter 4.2. Figure 4-2 is an example of the captured waveform of a 3Gbps signal. The waveform consists of 20 cycles of the 20-bit test pattern, with 400 samples in each data bit and 160,000 samples in total. This captured waveform is used to perform transmitter functional and parameter testing.

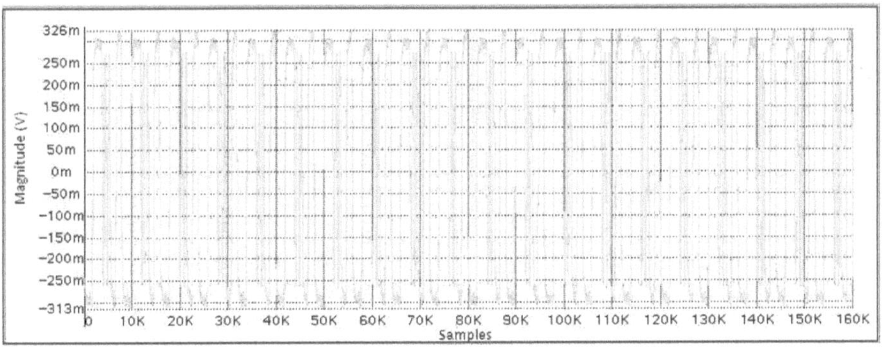

Fig. 4-2. Captured transmitter output signal

The original work in this chapter focuses on jitter extraction, as a key part of characterizing and testing the HSSI. Obviously, jitter is not the only parameter that needs to be characterized and tested in the HSSI. However, other measurements, such as transmitter function, rise/fall time and the amplitude characteristics, are straightforward to perform once we capture the waveform, and their test time is very short compared to the jitter testing.

4.3.1 Generating Edge Displacement

Basically, jitter is the edge displacement of the actual data edge transition position compared to its ideal position. In our test setup as shown in Figure 4-1, the reference clock *TX_ref_clk* determines the data rate of the transmitter. The transmitter ideal edge positions can be calculated by assuming that all the data bits are transmitted without any jitter. The actual position of each edge transition might deviate from its ideal position due to jitter.

Figure 4-3 shows an example of two edge transitions (*L* to *H* and *H* to *L*) captured using the digitizer. As the edge transitions contain various noise and interference sources in reality, they are not smooth functions of time. It is then suitable to use a curve fitting technique to extract actual zero crossing positions [103] in a way that reduces the effects of noise.

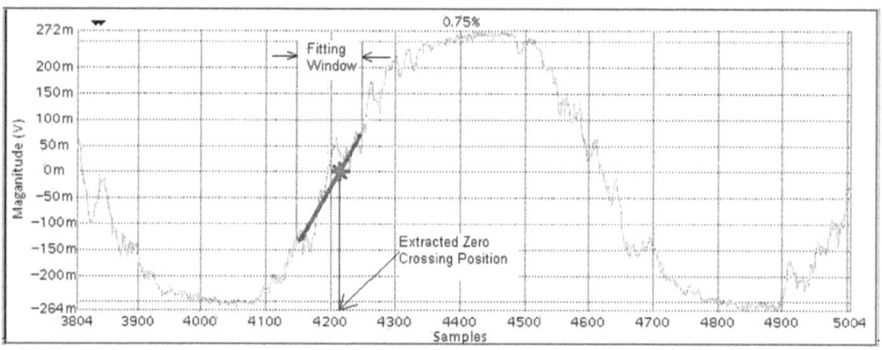

Fig. 4-3. Actual edge transitions and curve fitting

The curve fitting is done in a window centered in the edge transition period. According to the SATA specification [22], the transmitter rise/fall time (20% - 80%) of 3Gbps signals is between 0.2UI (67ps) and 0.41UI(136ps). When the effective sampling resolution is 400 samples per bit, the number of samples during an edge transition (20% - 80%) period is between 80 and 164. Therefore, we choose a window with 80 samples to perform the curve fitting as shown in Figure 4-3. The first captured zero crossing sample in a transition edge determines the centre of the window that we choose for the curve fitting.

Figure 4-4 plots all the edge transition positions calculated from our curve fitting technique. The *x*-axis denotes the edge sequence, which has 160 edges in the captured 400 data bits. The edge position in *y*-axis is denoted by the number of samples relative to the first edge.

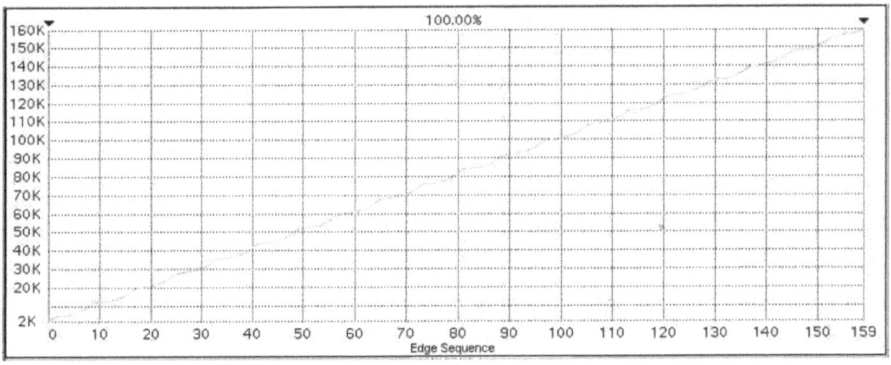

Fig. 4-4. Derived edge positions from curve fitting

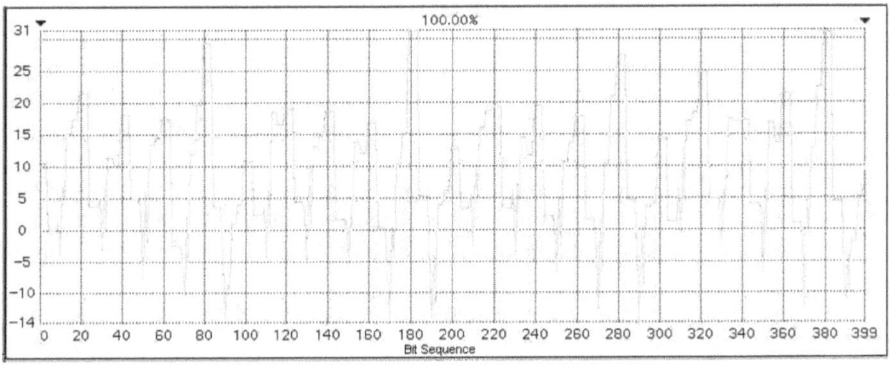

Fig. 4-5. Edge displacement data after interpolation

The edge displacement data is obtained from the derived edge position minus the ideal edge position. The ideal position is calculated based on the ideal data rate of the transmitter and the first derived edge position. In this way, we extract 160 samples of the edge displacement data from the 160 derived edge positions. To perform FFT for the jitter spectrum analysis, we need edge displacement information for every data bit.

In our implementation, we assume that no jitter is introduced in the data bits where no data transitions occur between two or more bits, so we just insert the edge displacement data from the previous edge transition to interpolate no-transition data bits. We will later develop the scheme to eliminate the effect that the interpolation may cause. Figure 4-5 illustrates the edge displacement data of all the 400 captured data bits. With the interpolation, the edge displacement data are equivalent to that obtained with a sampling rate of F_{DATA}, where F_{DATA} =3G for 3Gbps signals.

Once we get the edge displacement data, we can extract the DJ and RJ components based on their properties in both the time domain and the frequency domain. Then TJ can be obtained based on DJ and RJ. Extracting jitter from the time domain and the frequency domain is not a new topic and there are many existing solutions [92], [93], [104]. Our contribution in this area is that of having developed a fast and accurate test approach by sensibly setting test parameters for data acquisition and of proposing a jitter extraction procedure suitable in an ATE environment.

4.3.2 Time Domain Approach

In time domain, we can build in a straightforward manner the edge histograms of the test pattern to extract the RJ and DJ information for a device. The histograms are built by folding (overlaying) the extracted edge displacement data at a folding frequency

$$f_{fold} = \frac{F_{DATA}}{L_{pattern}} \tag{4-16}$$

In the 20-bit test pattern ($L_{pattern}$=20), there are eight transition edges. Figure 4-6 plots one cycle of the actually captured 20-bit test pattern as well as the ideal waveform.

Eight consecutive samples of the 160-sample edge displacement data (taken before interpolation) correspond to one cycle of the test pattern. We then obtain the actual edge histograms by folding the edge displacement data over every 8 samples.

The upper part of Figure 4-7 illustrates the histograms of the eight edges. We can obtain the mean value m_i and the SD value δ_i of each histogram. The mean values are shown in the lower part of Figure 4-7. The histogram information is used as a basis for extracting the main jitter components, such as RJ, DJ and TJ of the device.

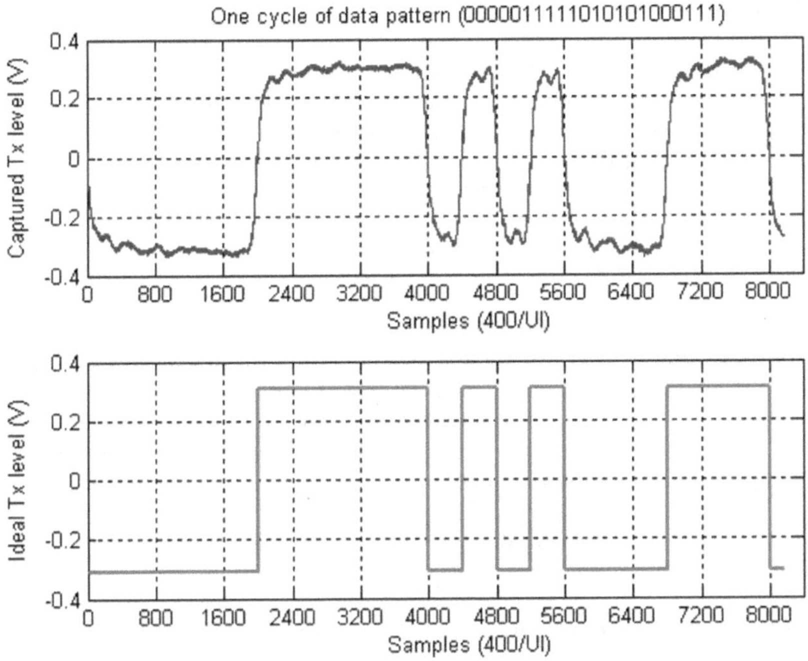

Fig. 4-6. One cycle of the test pattern

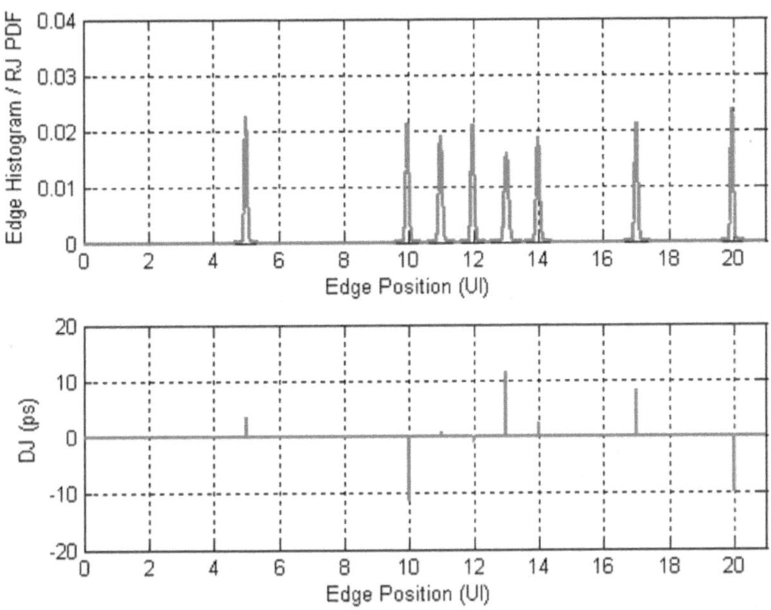

Fig. 4-7. Histograms and DJ of all eight edges

4.3.2.1 RJ Extraction

RJ is caused by random events, primarily by thermal noise in electrical components. As this kind of events exhibits a Gaussian distribution, we assume RJ is Gaussian [34], characterized by the SD value. The RJ value of the device is obtained by getting the RMS value of the SDs of the eight edge histograms:

$$RJ = \sqrt{\frac{\delta_1^2 + \delta_2^2 + ... + \delta_N^2}{N}}$$

where $N = 8$ for the 20-bit test pattern.

The RJ Gaussian property is also demonstrated by the actual edge histograms built from the captured data. As we can see from Figure 4-7 and Figure 4-12 (discussed later), the histograms are very close to Gaussian distributions even though there are only 20 samples in each edge histogram. Therefore, we represent the RJ at each edge using the Gaussian function

$$p(x) = \frac{1}{\delta\sqrt{2\pi}} e^{-(x)^2/2\delta^2}$$

where δ is the SD of the histogram at that edge. The RJ at each of the transition edges can lead us to get the TJ profile of the device once we get the DJ at each edge.

Because of the randomness of RJ, we need a lot of samples at each edge in order to capture the randomness in its histogram. Based on the analysis in Chapter 4.2, for the 20-bit test pattern, taking samples on 400 bits can achieve good accuracy (the statistics error is below 8%) within reasonable test time. The choice of capturing 400 bits is also proved to be reasonable by the fact that we still get similar jitter test results while we increase the number of captured bits.

4.3.2.2 DJ Extraction

By definition, the mean value m of an edge histogram would reflect the DJ at that edge, which gives

$$DJ_i = m_i$$

where i is the edge index.

The DJ of a device is the maximum value minus the minimum value of the DJ values at all edges, which gives

$$DJ = \max(DJ_1, DJ_2, ..., DJ_n) - \min(DJ_1, DJ_2, ..., DJ_n)$$

where n is the number of total edges. In our case, with the patter given, we have $n = 8$. The DJ value at each edge of the 20-bit data pattern is illustrated in the bottom part of Figure 4-7. The DJ of the device is then equal to the peak-to-peak value of the plot, which is 23.1ps, obtained as the DJ at the 13th UI minus the DJ at the 10th UI.

The lowest DJ frequency that can be cancelled and therefore excluded from the RJ is the folding frequency f_{fold}. Any DJ whose frequency is lower than f_{fold} will affect the RJ measurement accuracy. For the 20-bit test pattern ($L_{pattern}$=20) in 3Gbps applications (F_{DATA}=3G), according to Equation (4-16) we have f_{fold} = 150MHz. In our applications, the dominant fundamental DJ frequency is the word clock frequency (discussed in Chapter 4.3.3 and shown in Figure 4-9), which is same as the folding frequency. Therefore, we observe that the DJ components do not leak into RJ.

4.3.2.3 TJ Calculation

TJ is comprised of DJ and RJ. As RJ is unbounded, the TJ specification defined in any communication standard is actually the peak-to-peak value at a certain BER level. Different BER levels give different TJ peak-to-peak values and the characterization needs to account for that.

In order to extract the TJ peak-to-peak value, we need first to construct the TJ profile. As we already know the DJ and RJ profile at each transition edge of the data pattern, we can use that information to construct its TJ profile through convolution. Table 4-6 lists all the RJ and DJ values at each of the eight edges shown in Figure 4-7.

Table 4-6. RJ and DJ Values in Figure 4-7

Position	RJ RMS(ps)	DJ (ps)	Notes
Edge 1: 5th UI	1.64	3.5	
Edge 2: 10th UI	1.73	-11.4	Minimum DJ
Edge 3: 11th UI	1.95	0.7	
Edge 4: 12th UI	1.75	-0.8	
Edge 5: 13th UI	2.32	11.7	Maximum DJ
Edge 6: 14th UI	1.96	2.4	
Edge 7: 17th UI	1.73	8.4	
Edge 8: 20th UI	1.56	-9.9	

As discussed previously throughout this book, the RJ PDF at each edge can be characterized as

$$RJ\_PDF_i(x) = \frac{1}{\delta_i\sqrt{2\pi}} e^{-(x)^2/2\delta_i^2} \qquad (4\text{-}17)$$

where i is the edge index and δ_i is the RJ RMS at that edge.

As we know the actual DJ value at each edge, the TJ profile at an edge can be obtained by calculating the convolution between the values of RJ and DJ at that edge:

$$TJ\_PDF_i = RJ\_PDF_i \otimes DJ_i \tag{4-18}$$

where i is the edge index, i = 1,2,…,8. If we denote the DJ value at edge i with m_i, according to Equations (4-17) and (4-18), the TJ PDF at edge i is represented by

$$TJ\_PDF_i(x) = \frac{1}{\delta_i\sqrt{2\pi}} e^{-(x-m_i)^2/2\delta_i^2} \tag{4-19}$$

To be able to associate the TJ with the BER value, we now need to construct the Cumulative Distribution Function (CDF) of the TJ profile at each edge, obtained as:

$$TJ\_CDF_i(x) = \int_{-\infty}^{x} TJ\_PDF_i dx \tag{4-20}$$

The $TJ\_CDF_i(x)$ represents the probability that the jitter (please note that the jitter is defined as the edge displacement) resides within the range of $[-\infty, x]$. For a zero-mean Gaussian distribution, we have *CDF*(-∞)=0, *CDF*(0)=0.5 and *CDF*(∞)=1.

According to Equations (4-19) and (4-20), we have

$$TJ\_CDF_i(x) = \int_{-\infty}^{x} \frac{1}{\delta_i\sqrt{2\pi}} e^{-(x-m_i)^2/2\delta_i^2} dx$$

$$= 0.5 + 0.5 * erf(\frac{x - m_i}{\delta_i * \sqrt{2}}) \tag{4-21}$$

where *erf*(x) denotes the error function, defined as

$$erf(x) = \frac{2}{\sqrt{\pi}} \int_{0}^{x} e^{-t^2} dt$$

Once we get the TJ CDF at each edge, the TJ CDF of the device can be represented by

$$TJ\_CDF(x) = \frac{1}{8}\sum_{i=1}^{8} TJ\_CDF_i(x) \tag{4-22}$$

According to Equation (4-21), we can re-write Equation (4-22)

$$TJ\_CDF(x) = \frac{1}{8}\sum_{i=1}^{8}[0.5 + 0.5 * erf(\frac{x - m_i}{\delta_i * \sqrt{2}})] \qquad (4\text{-}23)$$

Figure 4-8 plots the PDF and CDF of the device TJ. According to the TJ CDF, we can get the TJ peak-to-peak value at a certain BER level by calculating the time difference between t_1 and t_2:

$$TJ_{peak-to-peak@BER} = t_2 - t_1 \qquad (4\text{-}24)$$

where t_1 and t_2 satisfy

$$TJ\_CDF(t_2) = 1 - BER/2$$

$$TJ\_CDF(t_1) = BER/2$$

For the TJ profile shown in Figure 4-8, according to Equation (4-24) we have

$$TJ_{pk2pk@10^{-12}} = 0.08275UI - (-0.06975UI)$$

$$= 0.15250UI$$

The above calculated TJ would reflect the TJ peak-to-peak value of the device at BER=10^{-12}. In the 3Gbps example run throughout this derivation, the calculated TJ value is equal to 50.8ps.

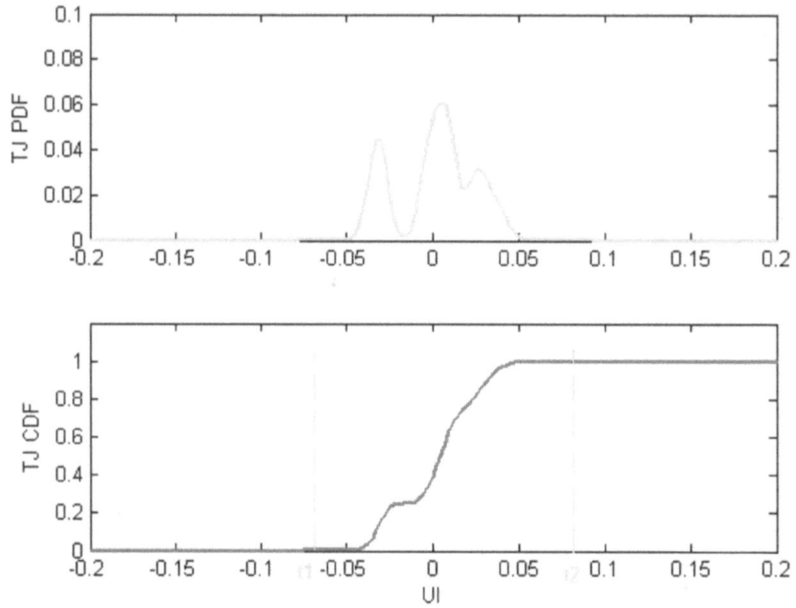

Fig. 4-8. The PDF and CDF of the device TJ

As we can see, the above TJ extraction process involves intensive computations and hence takes a lot of time. In production, we can estimate the RJ peak-to-peak value at a certain BER level by multiplying the RJ RMS value with the Q factor at that BER level. The TJ value can then be obtained by summing the DJ and the RJ peak-to-peak value:

$$TJ = DJ + 2 * Q(BER) * RJ_{RMS}$$

To obtain a tangible quantitative relation between the three types of jitter at the BER levels of most practical interest, we note that $Q(BER)$ is 7.035 at BER = 10^{-12} [34].

In the above example (with RJ and DJ values being as listed in Table 4-6), the TJ estimation based on the Q-factor gives a TJ value of 49.1ps. Quick comparison confirms that this value is very close to the TJ value calculated based on the TJ CDF profile (50.8ps). Therefore, it is fast and by far very much acceptable to use the developed Q factor-based method for TJ calculation in production testing scenarios.

4.3.3 Frequency Domain Approach

In the frequency domain, jitter components are extracted from the jitter spectrum. The TJ spectrum can be obtained by passing the edge displacement data as shown in Figure 4-5 through an FFT. Figure 4-9 illustrates the TJ spectrum of the captured signal shown in Figure 4-2. From the obtained spectrum, we can get the power at each frequency bin – we will denote each such power by Ci, where i is from 0 to 199.

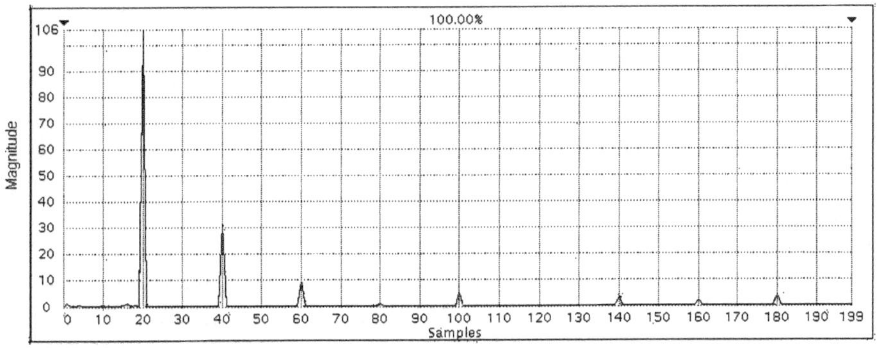

Fig. 4-9. TJ spectrum

4.3.3.1 RJ Extraction

In the TJ spectrum, RJ is the noise floor while DJ components are the impulses. The RJ RMS value is equivalent to the total noise power in the TJ spectrum. The noise power spectrum is constructed by replacing all the DJ frequency bins in the TJ spectrum with the average of the non-DJ frequency bins.

To remove the DJ completely for RJ extraction, this approach requires the DJ frequencies to be coherent [103]: all the DJ frequencies need to be exactly multiples of the FFT frequency resolution, as a non-coherent DJ frequency appears to consist of many frequency components in the FFT frequency bins and hence contaminates the RJ spectrum. In our applications, one DJ source is the device reference clock, which is 30MHz.

Another DJ source is the word clock of the device, which is 150MHz for 3G signals. The word clock is used in the HSSI to synchronize the parallel data. In addition, the ISI is also a DJ source. For the 20-bit data pattern, the ISI frequencies would be the multiples of 150MHz for 3G signals. For these facts and also according to the TJ spectrum, we know that all the DJ frequencies in our applications are multiples of 30MHz – the device reference clock frequency. In addition, as discussed in Chapter 4-2, our test setup strictly makes the reference clock frequency and the output data rate coherent. In our applications, the FFT frequency resolution is 7.5MHz for 3G signals.

Therefore, all the DJ frequencies are multiples of the FFT frequency resolution. Among the 200 frequency bins, 49 of them are DJ bins (all *Ci* with *i mod 4* = 0). To calculate the noise floor of the TJ spectrum, we replace all the DJ bins with the average of the RJ bins

$$C_{RJ\_average} = \frac{1}{150} * \left(\sum_{\substack{i=1 \\ i \bmod 4 \neq 0}}^{199} C_i \right)$$

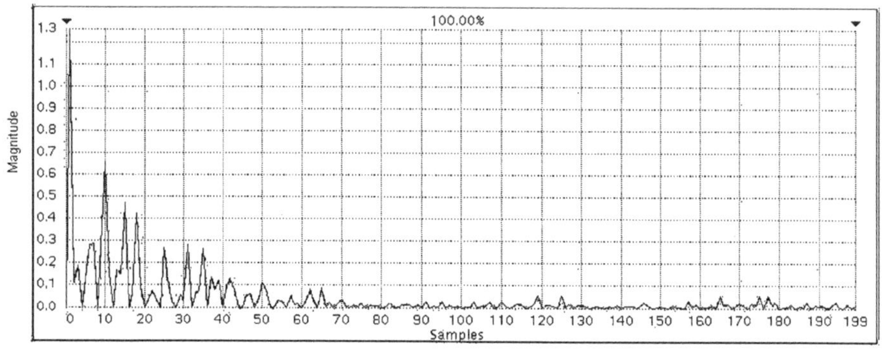

Fig. 4-10. RJ spectrum

Figure 4-10 plots the spectrum after the above replacement. It represents the RJ spectrum of the device. According to Parseval's theorem [103], the RMS value of the RJ spectrum is the square-root-of-sum-of-power of all bins given by

$$RJ = \sqrt{\sum_{\substack{k=1 \\ k \bmod 4 \neq 0}}^{199} C_k + 49 * C_{RJ\_average}}$$

4.3.3.2 DJ Extraction

In the frequency domain, we extract the device DJ component from its TJ spectrum. Similar to some commercial stand-along jitter equipment [104], we adopt the following steps for the DJ extraction:

(1) Obtaining the DJ-only spectrum by setting to zero all bins in the TJ spectrum that are attributable to RJ. In our case, we set to zero all the TJ bins that are not multiples of 4 (bin 4 corresponds to 30MHz)

(2) Performing an inverse FFT on the DJ-only spectrum to generate the time-domain data. The generated data would reflect the edge displacement that is only contributed by DJ.

(3) Getting the peak-to-peak value of the data excluding locations that actually do not have edge transitions. The peak-to-peak value is the DJ value of the device.

In step (3), we exclude the locations that actually do not have edge transitions when calculating the final DJ value. This would eliminate the artifacts that might have been introduced when we insert the edge displacement data on no-transition edges in order to perform the FFT.

Due to the DJ coherence constraint, we need to investigate the validity of each new design when using the spectrum approach for jitter extraction. One good thing is that the jitter spectrum is mainly determined by the device architecture (such as the CDR and PLL structure) and the test setup (the test hardware and the test pattern). Once a design is finalized, its jitter spectrum constitutes are constant. Therefore, the validation only needs to be done once for every new design, which makes it appealing.

4.3.4 Hybrid Approach

For standalone jitter testing equipment, the jitter is extracted on the background and the users have very limited control over the extraction process. The test results may vary from one instrument to another, depending on the jitter profiles in the test signal. For example, if there is uncorrelated DJ, some instruments may bin it to RJ and hence exaggerate RJ. Because we have the total control over the jitter extraction process, when needed, our hybrid approach can use both the time domain and the frequency domain data to provide a more accurate test result

As discussed previously, we can extract the jitter components from either the time domain or the frequency domain. Each approach has its advantages and disadvantages. If the test pattern length $L_{pattern}$ is small, such as 20, we prefer the time domain approach. The reasons are that we do not need to pay much attention to the actual DJ frequencies and that folding 20-bit data are not too complicated.

However, in some special cases, the limitation of the time domain approach may arise. As discussed in Chapter 4.3.2.2, the time domain approach cannot exclude DJ frequencies below the folding frequency from RJ. When we fold the data every 20 bits in 3Gbps applications, the lowest DJ frequency that can be excluded from RJ is 150MHz (F_{fold}). This normally does not cause issues in our applications because 150MHz and its multiples are the dominant DJ frequencies as shown in the TJ spectrum in Figure 4-9. However, we did observe that in some special cases there are DJ components with a frequency lower than F_{fold}. One example is the reference clock bleeding through.

The reference clock (30MHz in our applications) may bleed into the transmitter output through the loadboard ground or inside the device. Figure 4-11 captures in such a case the edge histograms that are folded at 150MHz but contain 30MHz DJ. In this case, the RJ distribution is not exactly Gaussian any more due to the DJ leakage. If we still use the SD to represent the RJ, the RJ would be exaggerated.

To exclude the 30MHz DJ frequency from RJ in the time domain, we need F_{fold} =30MHz. According to Equation (4-16), for 3Gbps applications we need the folding pattern length $L_{pattern}$=100. To build edge histograms, we need to capture at least 2000-bit data, assuming capturing 20 cycles.

In production testing of massively manufactured devices, we cannot afford the long test time required for the acquisition and the processing of such a large amount of data.

We are hence forced to explore the more practical alternatives. In our case, instead of lowering the folding frequency, we solve this problem by removing the low frequency DJ from the edge displacement data before building edge histograms.

We can actually set the specific low frequency DJ components in the jitter spectrum to zero and then perform an inverse FFT to get the edge displacement data that does not contain the low frequency DJ. Figure 4-12 plots an example of the histograms where the low frequency DJ has been removed in the manner that we just outlined.

After this removal step, the standard deviation of the histogram would much more closely reflect the true RJ of the device as the main source of the DJ spillover into the actual RJ (commonly referred to in industry as "bleeding") is at least attenuated, if not completely removed.

This approach will be also useful when we cannot get accurate RJ measurements using the frequency domain approach because DJ components are not coherent. We can use the hybrid approach to remove the uncorrelated jitter bins before calculating RJ.

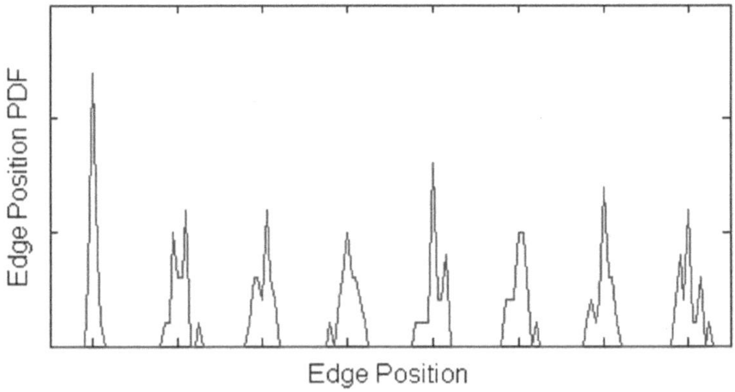

Fig. 4-11. Histograms with low frequency DJ: SD = 4.08Ps

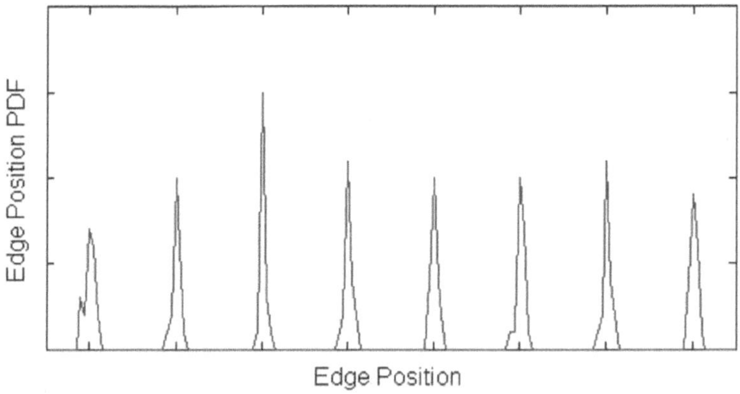

Fig. 4-12. Histograms after removing low frequency DJ: RJ = SD = 1.70Ps

4.3.5 Limitations of Each Approach

As already mentioned, there are advantages and disadvantages in each of the two jitter extraction approaches presented. It is useful to understand the advantages and the disadvantages well, to be able to use the right method for the application at hand.

To achieve a high accuracy, the frequency domain approach needs DJ components to be constant and also be coherent with the FFT frequency resolution. Our TJ spectrum demonstrates that this requirement is satisfied in our applications. However, in some applications, the DJ frequencies may not be coherent and even may vary from device to device. In this case, the frequency leakage may degrade the jitter test accuracy if we only rely on the frequency domain approach.

The time domain approach requires that the major DJ frequencies are multiples of the folding frequency in order to avoid DJ leakage. If we happen to have low frequency DJ, the folding frequency then must be also low. The time domain method can therefore require capture of a larger number of data bits and hence it will definitely require longer test time.

Finally, please note that although the hybrid approach can save some test time in the case of low frequency DJ, it still requires that the low frequency DJ components are consistent so we can effectively deal with it.

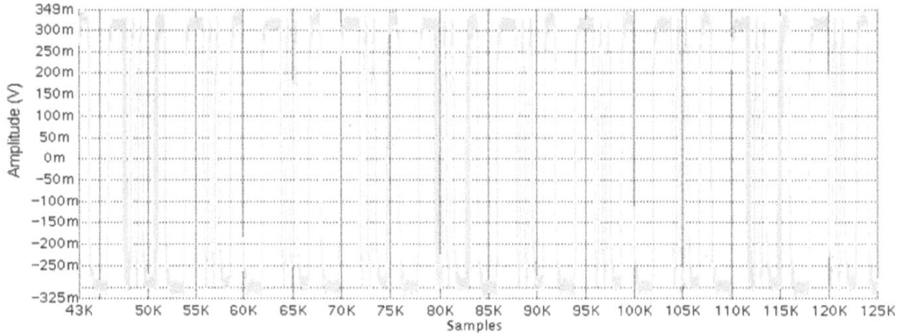

Fig. 4-13. DJ leakage

To detect the possible DJ leakage and uncorrelated DJ, we also implement a simple eye mask test. Figure 4-13 shows an example of captured waveform with DJ leakage. We calculate the eye mask by extracting the zero-crossing width of each edge and then overlay them together. The eye mask includes all kinds of jitter components, including non-Gaussian RJ.

4.4 Experimental Results

To evaluate a jitter test solution used in mass production, throughput and accuracy are the two most important criteria. We will now show that our experimental results demonstrate the superiority of our proposed solution in both throughput and accuracy.

In the ATE environment, every millisecond adds to the cost of the product. As discussed in Chapter 4-2, all the test parameters (pattern length, effective sampling rate, number of samples, and undersampling rate) in our solution have been optimized to keep the test time as short as possible while still capable of accurately capturing all the information we need for the transmitter tests. For both 3Gbps and 6Gbps applications, we managed to finish the entire transmitter testing procedure within only 100 milliseconds. The test time includes the data capture, the jitter extraction and other transmitter tests, such as transmitter function and rising/falling time tests.

Accuracy shows how close the measured jitter value is to its true value. The true value is usually obtained using a bench instrument whose accuracy has been verified and is widely accepted. Repeatability shows whether the test gives the same or similar result from run to run and from time to time for the same device while other conditions, such as supply voltages and temperature are the same. We have conducted intensive exploration of the repeatability and accuracy of our solution.

4.4.1 Bench Correlation

We have correlated our ATE jitter test results with the results obtained using the commercially available jitter test instrument Tektronix JIT3. Tektronix JIT3 is favored by many test/application engineers for its excellent jitter extraction ability and accuracy. Table 4-7 shows the results from 3 correlation devices. The ATE data in this table records the jitter mean values from the time domain approach with 20 runs for each device in a 3Gbps application. The repeatability of our ATE solution is discussed later.

Table 4-7. Jitter Measurement Results between ATE and Bench

Jitter/ Device	Device 1		Device 2		Device 3	
	Bench	ATE	Bench	ATE	Bench	ATE
RJ	1.9	1.92	2.05	1.83	2.02	1.81
DJ	19.9	21.5	27.3	29.4	25	26.5
TJ	40.8	48.38	49.3	55.02	45.8	51.84

As we can see, the RJ difference between the bench and ATE is within 0.2ps; the DJ difference is within 3ps (DJ from ATE is consistently slightly higher than that from the bench equipment). As we know, absolute correlation in numbers for different jitter test solutions rarely happens. Considering this is done on ATE with a completely different instrument and setup from the bench environment, the correlation result is very good.

One reason for the higher DJ on ATE is that the signal path on ATE is longer than that on bench. The longer signal path can introduce more ISI, and hence results in higher DJ on ATE. In addition, the different TJ extrapolation algorithms between the bench and ATE also introduce difference in the final TJ report.

In Table 4-7, the three correlation devices generate similar amounts of jitter. We further conduct the correlation between ATE and bench equipment using an alternative reference clock to generate higher RJ and DJ. More details and correlation data are presented in Chapter 4.4.4.

4.4.2 Correlating Two RJ Approaches

As we have discussed, the RJ can be extracted from both the time domain and the frequency domain. The frequency domain approach is less pattern-dependent as it does not involve building histograms. This approach is preferred on ATE if we need to investigate the jitter performance with different test patterns. However, the results from this approach need to be verified as the extraction process involves data interpolation and jitter component replacement. These steps might introduce errors as the assumptions for these steps may become invalid at a certain condition. In addition, the frequency domain approach requires that the DJ frequency leakage is negligible, which may not be satisfied in some cases.

On the other hand, the time-domain approach is very straightforward. It can be used to correlate the test results from the frequency domain. Figure 4-14 shows the test results of a device with 20 runs on ATE, where *RJ_Spectrum* is the RJ value from the frequency domain approach, and *RJ_Timing* is the RJ value from the time domain approach. It demonstrates that both approaches exhibit good repeatability and the correlation is also very good.

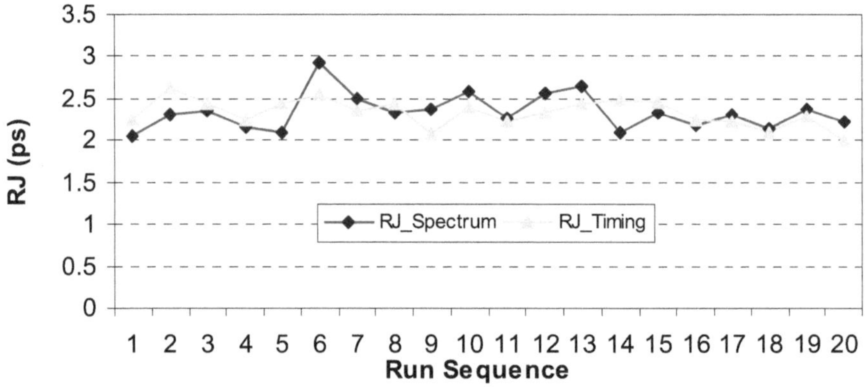

Fig. 4-14. RJ repeatability and correlation

We also evaluated the correlation between the two approaches across PVT corners. Figure 4-15 plots the jitter distribution across the PVT corners from both approaches, where the x-axis denotes the measured RJ values and the y-axis represents the number of hits. The difference between the two approaches is very small: the measured jitter mean difference is only 0.2ps and distribution profiles are very similar.

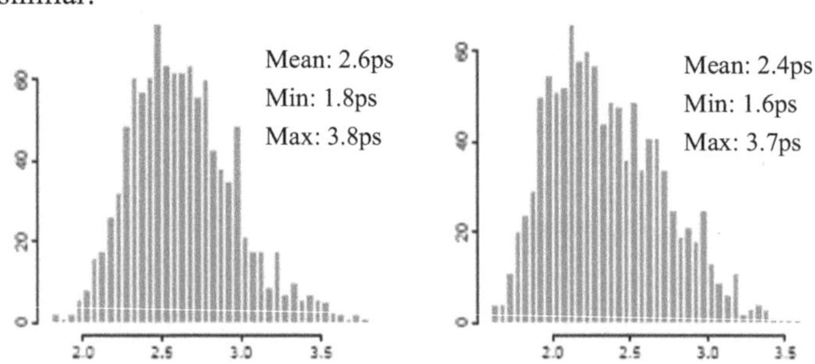

(a) RJ_Spectrum distribution (b) RJ_Timing distribution

Fig. 4-15. Jitter distribution across PVT corners

As we can see, the device-to-device correlation between the bench and the ATE time domain approach is very good. On ATE, the time domain and frequency domain approaches also correlate well from either multiple runs for a single device or a larger number of devices across all conditions. These experiments confirm the claim of the excellent accuracy and repeatability of our proposed jitter test procedure.

4.4.3 Impact of Test Patterns

In our testing procedures presented so far, we use a 20-bit test pattern. Even though the test pattern includes both high transition density and low transition density data, the length of the pattern is short and the actual data in real applications has more variations. In order to make sure that the test results using the 20-bit test pattern actually can represent the performance of the device in a real environment, we now need to investigate the test pattern impact to jitter measurement results.

Table 4-8 lists the jitter measurement results using the 20-bit test pattern, the 128-bit PRBS pattern discussed in Chapter 3 and a clock pattern (101010..) respectively.

Table 4-8. Jitter Measurement Results Using Different Test Patterns

Jitter/Pattern	20-bit	PRBS	1010
RJ	2.07	1.99	2.25
DJ	41.3	48.8	34.5
PJ	26.3	30.25	23.1
DCD	1.09	0.84	2.37
ISI	13.9	17.75	9.37
TJ	60.8	65.34	56.5

We take the measurements from an ATE loadboard. Instead of connecting the transmitter output to the digitizer on ATE, we connect the transmitter outputs to the bench instrument JIT3 to investigate the impacts of the test patterns. There are three main benefits of this approach:

1) It adopts the real production environment and hence can directly use the results to evaluate our ATE solution
2) Measurement results done on the "bench" equipment in general are more acceptable and give more confidence level to the overall experiment results
3) There is no need to develop ATE test program for new test patterns.

As we can see, the jitter numbers from all the three test patterns are very close. The 20-bit test pattern generates a similar RJ value compared to the 128-bit PRBS pattern or the clock pattern. It also generates moderate DJ: slightly lower (less than 5ps) than the PRBS pattern and slightly higher than the clock pattern (the difference is caused by ISI). Therefore, the jitter measurements from the 20-bit test pattern we used provide a good representation of the true jitter performance of the device.

4.4.4 Impact of the Reference Clock

As shown in Figure 4-1 -- Transmitter Test Setup for Data Acquisition, the HSD on ATE provides a reference clock to the transmitter PLL. The transmitter output data is then synchronized by an internal transmitter clock generated by the PLL. The quality of the reference clock can affect the jitter in the transmitter output signal.

To investigate the reference clock impact to the final transmitter jitter numbers, instead of using the HSD clock, we create another clock using the AWG6000 on the ATE as the reference clock to the transmitter. The AWG clock is generated by directly storing consecutive zeros and ones in the AWG according to the transmitter reference clock frequency requirement. We then measure the transmitter jitter using the two different reference clocks. Table 4-9 shows the measurement results using the bench equipment JIT3 and our jitter extraction techniques on ATE.

Table 4-9. Reference Clock Impacts to Transmitter Jitter Measurement

Jitter / Device	HSD clock		AWG clock	
	Bench	ATE	Bench	ATE
RJ	2.07	2.3	5.7	5.9
DJ	41.3	29.07	50.6	41.06
TJ	60.8	67.0	112.1	127.8

As we can see, using the reference clock generated by AWG (AWG clock), the transmitter exhibits higher jitter than using the reference clock generated by an HSD channel (HSD clock). One reason is that the AWG has much higher bandwidth than the HSD channel, and hence more high frequency jitter in the AWG clock transfers into the transmitter clock. Table 4-9 also further verifies our ATE jitter extraction techniques: the ATE results correlate with the bench results under different jitter levels.

The intrinsic jitter in reference clocks can be reduced by a clever design. A method is presented in [130] to reduce jitter in clock signals at the cost of extra hardware. It uses real-time averaging to combine multiple signals in order to produce one output signal with much lower random jitter. The reduced jitter can be expressed theoretically by

$$\delta_{output} = \frac{\sqrt{\delta_1^2 + \delta_2^2 + \ldots + \delta_N^2}}{N}$$

where δ_{output} is the jitter of the output signal, N is the number of combined signals and δ_i (i=1, 2, …, N) is their jitter. The experimental result in [130] shows a 3x reduction in jitter (from 4ps to 1.3ps) by combing eight signals. This is very close to the above theoretical analysis. Interestingly, similar to our CTL Gaussian generator from Chapter 6.2.2.3, it exploits the central limit theorem.

Even though a better reference clock can help report better transmitter jitter numbers, in production we need to try to mimic the real application environment, where the reference clock is usually provided by a crystal oscillator. We correlate the ATE result (the reference clock is provided by an HSD on ATE) with the bench evaluation board result where the reference clock is generated by a crystal oscillator, and the results are very close.

4.4.5 Extending to 6 Gbps Applications

Although our previous discussion and experiments concentrate on 3Gbps applications, our approach applies to any data rates as long as the digitizer bandwidth is sufficient. Because the bandwidth of our ATE instrument is now above 9 GHz, we can extend our transmitter jitter test solution from 3Gbps applications to 6Gbps applications.

For 6Gbps applications, we only need to adjust the reference clock and the under-sampling clock, according to the data acquisition principles discussed in Chapter 4-2 and the test setup parameters listed in Table 4-4. Figure 4-16 shows a part of the 6Gbps waveform captured using our solution.

Based on the obtained waveform, we can then extract the jitter components using exactly the same scheme as discussed for 3G signals. Figure 4-17 shows the measured jitter at 6Gbps data rate from one device with 20 runs, where the upper part plots both DJ and RJ and the lower part plots RJ only.

At 6Gbps data rate, our solution still provides similar accuracy and repeatability to that at 3G applications. As shown in Figure 4-17, the RJ variation from run to run is within ±0.5ps. The measured jitter on ATE also correlates well with bench results.

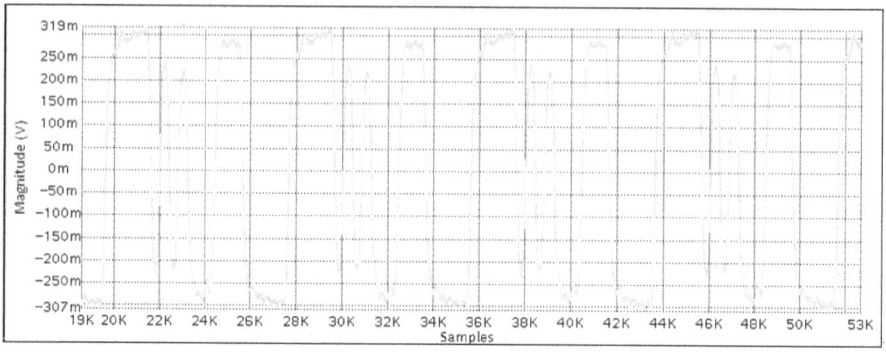

Fig. 4-16. Captured 6G waveform (only 45 bits shown)

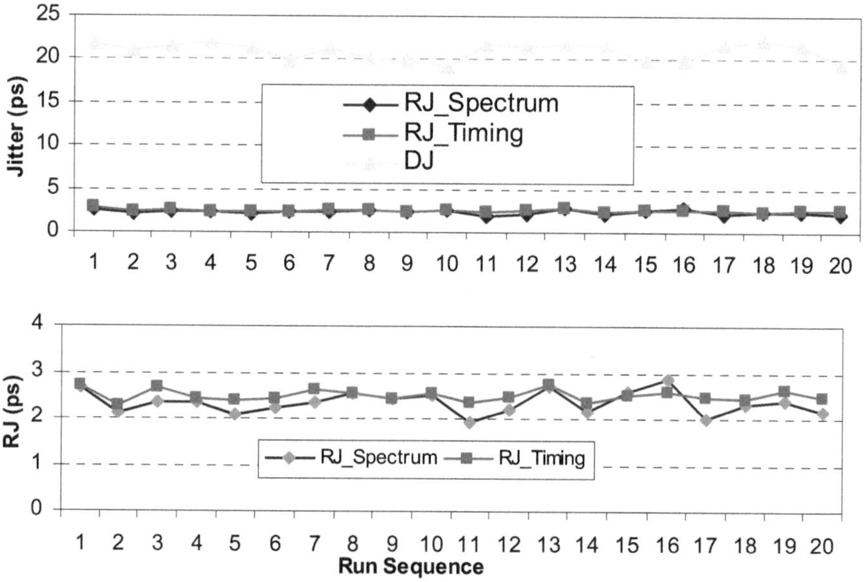

Fig. 4-17. RJ and DJ at 6G data rate

4.5 Summary

In this chapter, we have presented a systematic approach to extract transmitter jitter. By just capturing an adequate number of samples with adequate resolution, we can achieve the whole transmitter testing within 100ms while keeping sub picosecond jitter accuracy. We make it feasible to run the time-consuming jitter test in an ATE environment with data rates up to 6Gbps.

To extract jitter, we first obtain the edge displacement data by comparing the actual zero-crossing points with their ideal positions. Based on that edge displacement information, we can easily extract the jitter characteristics from either the time domain or the frequency domain. To guarantee accuracy while keeping test time short, we further proposed a hybrid approach. Its main benefit is that it prevents low frequency DJ from "bleeding" into RJ. Experimental data demonstrates the validity of our approaches, as well as the ability to easily scale up to higher frequencies.

5 Testing HSSIs with or without ATE Instruments

Abstract *This chapter introduces techniques that do not relay on high-speed ATE instruments for both the transmitter and the receiver testing that is discussed in Chapter 3 and Chapter 4, respectively. We describe the principles of the design of FPGA-based test equipment, including Bit Error Rate Tester (BERT), pseudo-random noise injection and channel emulation. Then, we provide the details of a complete standalone tester that uses relays and/or MEMS-based switching devices. The advantages and disadvantages of the state-of-the art in each such case are presented.*

As we discussed in Chapter 3 and Chapter 4, the ATE-based solutions greatly speed-up the receiver jitter tolerance and transmitter jitter qualifications. However, there are two major limitations in applying the ATE-based approaches for HSSI qualifications.

The first one is the number of ATE instruments available. For each HSSI, we need an AWG and a digitizer in order to do parallel testing. Nowadays, there are some devices with a few tens or even more than 100 HSSIs. No ATE platform can accommodate so many AWGs and Digitizers. In addition, the AWGs and Digitizers that can handle multiple GHz signals are very expensive. The AWG/Digitizer approach is prohibitive from the cost point of view for devices with multiple HSSIs.

The second limitation is the lack of high-speed instruments for the latest HSSI receiver testing. AWG6000 is the best AWG on ATE from what we know in terms of speed, but it can only perform jitter tolerance testing for devices with data rates up to 3Gbps. Several ATE suppliers have provided production pin-card solutions up to 6Gbps. However, for higher speed applications, such as 10Gbps and above, systematic ATE solutions are not mature yet, even though there is lots of ongoing research in that direction, e.g., [131].

In any case, because ATE equipment is very expensive and it takes long time to amortize it in normal use, it is a fair assumption that the performance of ATE instruments used in production testing will be lagging behind the performance of the devices developed using the state-of-the-art technology.

To overcome these limitations, this chapter presents HSSI testing techniques that do not necessarily rely on ATE instruments. We first introduce common DFT techniques in HSSIs. We then present our solutions that do not rely on ATE in-

Y. Fan, Z. Zilic, *Accelerating Test, Validation and Debug of High Speed Serial Interfaces*,
DOI 10.1007/978-90-481-9398-1_5,

struments and DFT techniques. We present an FPGA-based bit error detection scheme for HSSI function validation and testing. We also propose a low-cost loopback-based testing scheme, where a novel jitter injection method is proposed using the latest phase delay lines. The loopback scheme can be applied to test HSSIs with data rates up to 12.5Gbps. It is also suitable for multi-lane HSSI testing. By using state-of-the-art high-speed relays, we combine the ATE solutions and the loopback solution along with the FPGA-based BERT to provide a more versatile scheme for HSSI validation and testing.

5.1 DFT in HSSIs

In order to qualify HSSIs, we need the speed of the test instruments to be higher than the data rate of the DUT. While the HSSI data rate keeps increasing, the availability of high-speed ATE instrument is always a bottleneck in HSSI testing. In addition, the test cost is very high even there are high-speed instruments available.

A mixed-signal tester capable of testing Gigabit serial interfaces normally costs millions of dollars. Many DFT techniques have been developed to ease the instrument requirement and reduce test cost.

5.1.1 Internal BERT

A BERT consists of a pattern generator and an error detector. The pattern generator can generate one or several patterns. The pattern can be used as the input of the transmitter.

The error detector can detect bit errors of the receiver after the received pattern is aligned and synchronized. The pattern generator and the error detector are usually used together. One such application is based on driving the transmitter using the pattern generator, sending the transmitter output to the input of the receiver, and then using the error detector in the receiver to check for any bit errors generated in the process.

The pattern generator and the error detector can also been used separately. As discussed in Chapter 3.3.2, a DFT-based error detector is used to check the bit errors of the receiver while the pattern generator is not used; the input signal of the receiver is provided by a high-speed ATE instrument - AWG. The cost of implementing of a BERT is relatively low. Chapter 5.2 will give an example of the detailed implementations.

5.1.2 Internal Loopback

In Chapter 3, we introduce techniques to test the receiver using a high-speed AWG as the signal generator; for the transmitter testing techniques presented in Chapter 4, we employ a high-bandwidth digitizer to capture the transmitter output, from which the functionality is checked and the transmitter parameters are extracted. These approaches relay on high-end test instruments. By looping the transmitter output back to the receiver input either internal or external, loopback provides mechanisms to test an HSSI without the need of high-speed instruments. Internal loopback is one of the most popular DFT techniques. Loopback testing has been widely used to check the functionality of HSSIs [69], [111].

Internal loopback is implemented by looping internal nodes of the transmitter to corresponding nodes in the receiver inside the HSSI. According to where the signal is looped back, different loopback modes can be built. Figure 5-1 illustrates major internal loopback paths:

Path 1 -- Near-end loopback mode, where the parallel data from the transmitter digital logic is returned back to the digital logic after the deserializer in the receiver.

Path 2 -- Far-end loopback mode 1, where the transmitter pre-driver output is fed back into the input of the deserializer.

Path 3 – Far-end loopback mode 2, where the transmitter driver output is fed back to the receiver input.

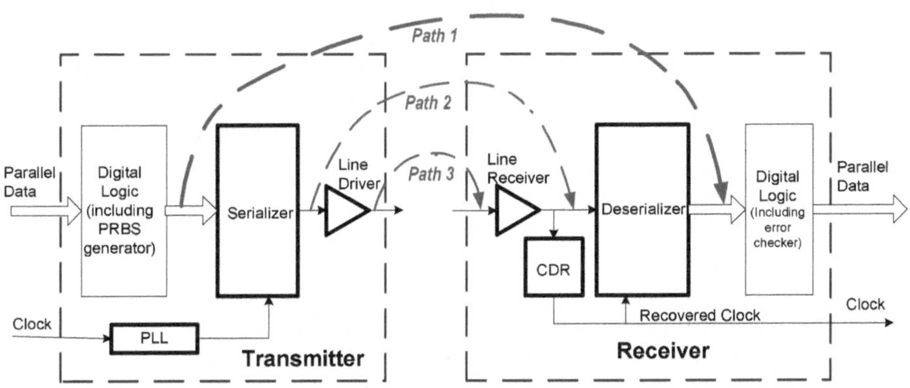

Fig. 5-1. Internal Loopback modes

Internal loopback test paths can be constructed using a multiplex/de-multiplex. As shown in Figure 5-2, a Mux can be used to choose the input signal from either the normal signal or the test signal. The signal generated by the Pattern Generator shown in Figure 3-27 is an example of the test signal. A De-Mux can be used to send the output to either the normal signal path or the test signal path.

Examples of test signal outputs include the signal paths to the Error Counter shown in Figure 3-27, Path1, Path 2 and Path 3 shown in Figure 5-1. The Mux and De-Mux are usually controlled independently in order to provide more flexibility in testing configuration.

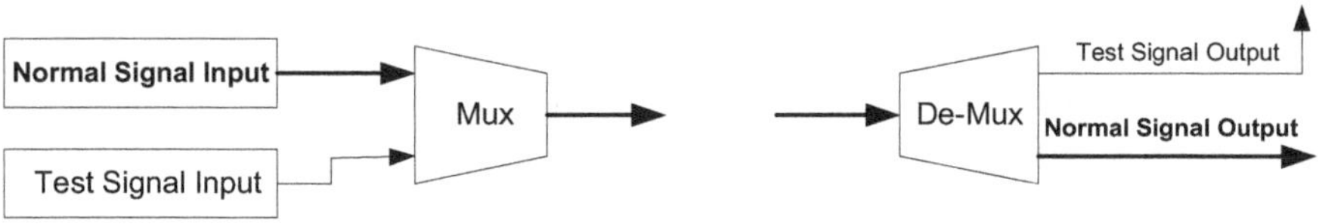

Fig. 5-2. Mux and De-Mux

One possible issue associated with the Mux/De-mux approach is that the extra circuitry may degrade the performance of the normal signal path, especially in the high-speed serial data path. One alternative that can reduce/avoid the disturbance is to tap the test signal from the normal signal path in a high impedance manner, such as a big resistor. In this way, the signal in the test path might be weaker than the signal in the normal path, but the integrity of the test signal is usually still good for debugging and functionality verification.

5.1.3 Other DFT Techniques

As we are keeping pushing up the data rate, it is critical in verifying that the circuit performance of each sub-block in an HSSI meets expectations. As shown in Fig. 2-4, an HSSI may only have a few pins for parallel data in/out, serial data in/out and a reference clock. To debug and test internal blocks, such as the serializer and the line driver, we need observability to critical internal nodes. Implementing test buses, including analog and/or digital ones, is a DFT technique to observe the internal signals. Under this technique, the internal signals are muxed to test buses and the test buses are de-muxed to device pins in test modes.

Some examples of critical internal signals include the transmitter and receiver clocks, analog front end outputs, and equalizer outputs. The test buses are also used to verify analog voltages/currents, such as critical bias points, bandgap reference voltages and loop filter voltages. The observed signals/voltages are very useful during the debugging/root cause analysis and when verifying the devices across a large sample of parts.

Boundary scan is a test technique that we can readily use to initiate the DFT functions, such as setting the loopback mode and controlling what internal signals are brought out to the test bus. Boundary scan, also known as Joint Test Action Group (JTAG) or IEEE 1149, was introduced in early 1990s and is now widely used for the test of electronic devices/systems at all stages of their lifecycle, as well as for the debug and additional functionality such as in-system reprogramming.

IEEE 1149.6 standardizes the boundary scan structures and methods for advanced digital networks, especially for those that are AC-coupled, differential or both [201]. The IEEE 1149.6 is suitable for HSSIs [202] and is hence widely implemented in HSSI to facilitate testing and debugging.

In recent years, more DFT techniques have emerged to detect certain failure mechanisms and process dependencies of devices. S. Sunter *et al.* proposed an undersampling BIST technique in [44] to measure parameters that affect jitter tolerance in multi-Gbps serial interfaces. The technique requires the receiver to use a reference clock that is slightly frequency-offset relative to the transmitter's reference clock in a loopback mode. Using undersampling, an Unlimited Time Resolution Analysis module is used to extract high-frequency jitter, transition-density dependent phase-shift, mean sampling position, sampling clock phase error and pin-to-pin skew. The module can be implemented on chip without changing the HSSI macro.

K-L Kim *et al.* proposes a duty cycle jitter measurement BIST in [200] to evaluate the duty cycle distribution of a clock or data with alternating bit sequence. The duty cycle distribution of the signal can be observed from the generated histograms.

5.1.4 Limitations of DFTs

Even though DFT is becoming increasingly important in HSSIs, DFT is not a solution for all. There are limitations associated with DFTs. First, the added circuitry can degrade the performance of the device. Because we are pushing the speed envelop in order to get more bandwidth, many HSSI devices have very little performance margin. Very small performance degradation might be problematic.

In addition, almost each DFT technique has its own limitations. For example, for the widely used loopback testing technique, the transmitter and the receiver work in synchronous mode in the loopback configuration, which is different from in-field asynchronous applications; therefore, process variation and defect mechanisms that affect both the transmitter and receiver might get masked

5.2 FPGA-based Bit Error Detection

In Chapter 3.3, we present two mechanisms to detect bit errors. The ATE based solution utilizes high-speed digital channels on ATE to compare the receiver recovered parallel data with the expected data. There are two limitations for this approach. First, it is very difficult to achieve synchronization between the DUT and ATE -- it depends on some special macros from the ATE vendor. In addition, it requires many high-speed digital channels, which are expensive and sometimes might not be available. Even though the DFT-based approach almost does not need any ATE instruments, it needs extra silicon area and design effort to implement and has to be planned before the HSSI is designed. In addition, once the design is finished, we can only choose the patterns that have been designed into the DFT; we do not have the ability to program new patterns.

To overcome these limitations, we explore FPGA-based approaches. In recent years, FPGAs have been widely used in testing applications because of their strong capabilities in generating test patterns, providing high-speed interfaces and performing configurable user-defined functions [53], [96], [135], [138]. One important FPGA application is Built-Off-Self-Test (BOST), where the test or BIST circuit is implemented in an FPGA that is placed off the chip on the test fixture [105].

One example of the BOST approach is presented by Sunter and Roy in [72], where an FPGA on a DUT interface board is used to test HSSIs. This BOST approach implements three BIST techniques [44], [106], [107] in the FPGA. In this section, we introduce a BOST approach for bit error detection [6]. Implemented in an FPGA, our approach performs similar function as the DFT feature discussed in Chapter 3.3.2 does, but the user has the freedom to set the test pattern. The solution does not need any ATE instrument or any DFT feature, and can be used to test almost any HSSI.

As discussed in Chapter 2.1.2, the basic concept of BER measurement is as follows: the pattern generator sends a data stream to a DUT; the error detector conducts a bit-by-bit comparison of the received signal from the DUT and records bit errors. According to applications, a BER tester (BERT) can be either serial or parallel.

5.2.1 Implementing a Serial BERT

A serial BERT sends serial bit sequences to a DUT and evaluates the output from the DUT. The DUT can be any serial digital communication link. The structure of a serial BERT is proposed and shown in Figure 5-3. In this scheme, the shift register shift_reg1 and the gate XOR1 form a LFSR. As the pattern generator of the serial BERT, the LFSR generates Pseudo Random Bit Sequences (PRBSs). These sequences are then sent to the DUT.

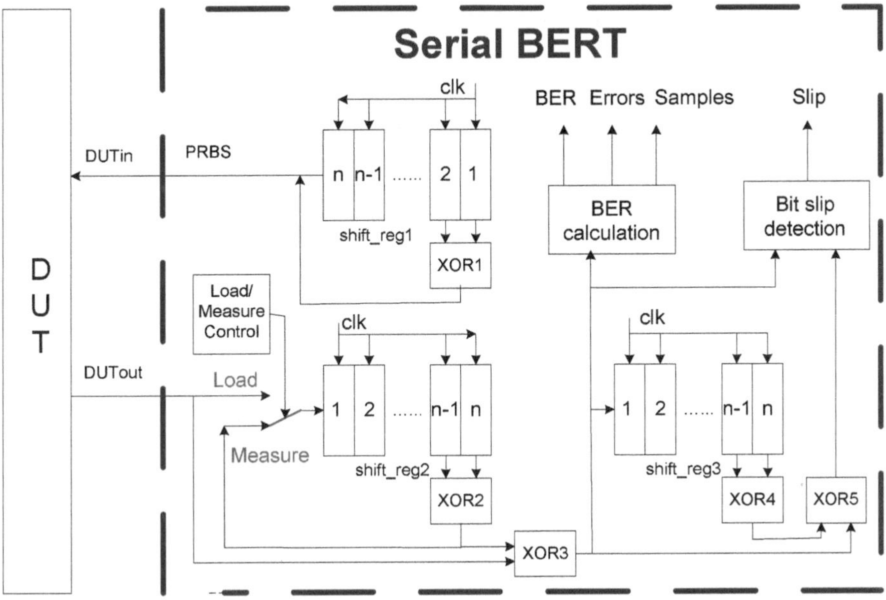

Fig. 5-3. Block diagram of a serial BERT

Before a measurement begins, the load/measure switch is set to be in *load* state until the shift_reg2 is full loaded with the contents of the shift_reg1. The switch is changed to *measure* state to start the BER measurement. The shift register shift_reg2, the switch and the gate XOR2 are used for synchronization. They generate a reference pattern by replicating the PRBS from the shift_reg1, but delaying the phase.

During the synchronization process, it is assumed that all the bits are correctly transmitted. The gate XOR3 serves as a comparator. It compares the pattern from the DUT to the reference pattern. If the test pattern is correctly transmitted by the DUT, then the two inputs of XOR3 should be the same value in each clock cycle. In a real BER measurement, the output of XOR3 is monitored every clock cycle: if a '1' is detected, a transmission error is counted; otherwise, the transmission is error-free.

In a real communication system, the transmission errors are in forms of a single-bit error, error bursts or bit slips. Bit slips result from a bit loss or a bit repeat. If a bit slip happens, only the repeated or lost bits should be counted as errors. We employ a mechanism to distinguish between error bursts and bit slips and to eliminate false long-term errors. In Figure 5-3, the shift register shift_reg3 and the gates XOR4 and XOR5 perform bit slip detection. The solution is based on the fact that the addition or superimposition of two PRBSs shifted in phase relative to each other produces another PRBS [108]. By monitoring the output of XOR3 and XOR5, it can be determined whether a bit slip happens.

5.2.2 Implementing a Parallel BERT

A parallel BERT is used to test communication interfaces that transmit parallel data. The implementation of the parallel BERT is based on the serial BERT. We build a k-bit parallel BERT using k independent serial BERTs, where k is the width of the parallel data (bit0 ~ bit(k-1)).

The parallel BERT sends Pseudo Random Word Sequences (PRWSs) to the DUT. In order to achieve randomness in the generated sequences, the independence of each serial BERT is important. Therefore, the lengths of the shift registers in the serial BERTs should be different in order to be able to have different periods [109].

We need to remove the circuit redundancy when k independent serial BERTs are directly put together to implement a parallel BERT. Each of the serial BERTs has circuits for the *load/measure* switch control and bit slip detection. However, the load/measure switches for the parallel data bits should change the state at the same time. Therefore, only one of the k such control circuits is needed for the switch control and word slip detection. Figure 5-4 shows the structure of the proposed parallel BERT. In the design, the serial BERT control circuitry for bit0 is used for the load/measure switch control and the word slip detection.

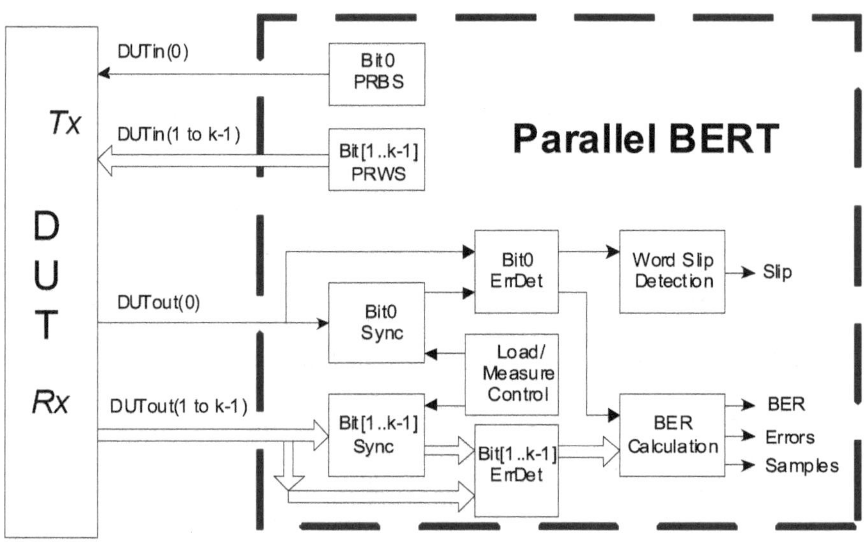

Fig. 5-4. Block diagram of the parallel BERT

5.2.3. HSSI Testing Demonstration

The BERT design has been built in VHDL, and can target almost any FPGA devices. To demonstrate the functionality of the BERT, we use it to test the HSSI in the Altera Mercury FPGA EP1M120F484C7 [10]. The Mercury HSSI can transmit and receive high-speed serial data streams with speed up to 1.25Gbps. Figure 5-5 shows the setup used to test the HSSI.

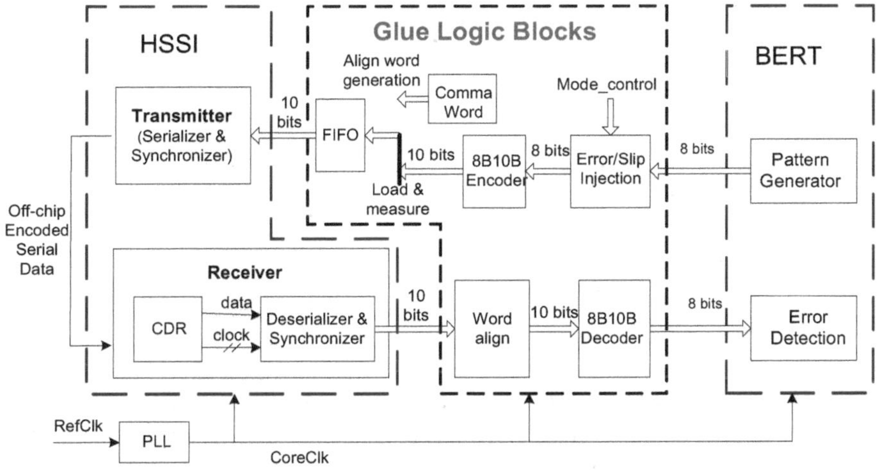

Fig. 5-5. HSSI testing setup to verify BERT functionality

In the testing setup, the HSSI is built by instantiating the Altera Mercury Gigabit Transceiver MegaCore [110]. The data width of the BERT is 8 bits. The glue logic is developed to directly interface the BERT and the HSSI. The Error/Slip Injection block inserts errors or word slips and could be used to test/verify the BERT.

The 8B10B encoder encodes the 8-bit sequences to 10-bit sequences to ensure enough bit transitions in the serial link for date recovery as discussed in Chapter 2.1.1. A FIFO is used to ensure that there is always data ready for transmission after a testing begins. Comma words are inserted at the start of the testing for word alignment. The 8B10B decoder recovers the 8-bit PRWSs sent by the BERT.

The whole testing setup (the HSSI, BERT and the glue logic blocks) is implemented in VHDL, targeting the EP1M120F484C7 device using Quartus II software. The synthesized results are downloaded onto an Altera Mercury CDR Demo board. The outputs of the transmitter are connected to the inputs of the receiver by two Sub-Miniature type-A (SMA) cables. In this setup, the data signal is running at 1.25Gbps. Higher data rates can be realized using higher performance FPGAs.

We obtained zero BER both from simulations and from running real tests on the board when no error or bit slip was injected. We also captured the transmitter output signal using an oscilloscope and verified the sequence was what we expected.

When the slip was injected on the test board, the Error Detector detected bit errors and the slip output was asserted. When only bit errors were injected, the BERT reported bit errors but bit slip output was not asserted. The experiments that we have performed demonstrate that even the relatively low-performance and inexpensive Altera Mercury HSSI can successfully serialize the parallel data, transmit the high-speed serial data over the cables, and recover the serial data sequence back to the original parallel data. The experiment also verifies the FPGA-based bit error detection scheme. Further experiments using the BERT will be demonstrated in Chapter 6-3.

5.3 Loopback Testing with Jitter Injection

5.3.1 Testing Setup

In Chapter 3, we present an ATE-based HSSI receiver testing scheme, where controllable amounts of jitter are injected through the AWG. This approach can successfully test the jitter tolerance performance for data rates up to 3Gbps. For data rates above 3Gbps, we cannot use the AWG-based jitter injection approach because the maximum sampling rate of the AWG is limited to 6Giga samples per second.

To overcome the ATE instrument limitation, we propose a loopback testing scheme. By looping the transmitter output back to the receiver input either internal or external, loopback testing has been widely used to check the functionality of HSSIs [69], [111].

The HSSI testing demonstrated in Chapter 5.2.3 is one example of the external loopback. Traditional loopback approaches do not have the capability to qualify design parameters, such as transmitter jitter and receiver jitter tolerance. Recent research shows the directions that use external loopback to verify design parameters. However, they need either special DFT features or extra special instrument modules [98], [99].

Our approach does not rely on any DFT features or special instruments; it only needs a few extra components that can fit into a testing loadboard. The approach is especially attractive for testing multiple-lane HSSIs or HSSIs with data rates above 6Gbps, where mature ATE solutions are not available. Figure 5-6 shows the block diagram of the proposed loopback testing. With this approach, we inject a controllable amount of jitter to the output of the transmitter signal using a phase delay line, and then feed the signal back (the operation is often denoted in practice with a verb "to loop back") to the input of the receiver. When the injected jitter is below a certain level, the receiver should be able to recover the transmitted data.

Otherwise, the device is defective. The details of the phase delay based jitter injection are discussed in Chapter 5.3.2.

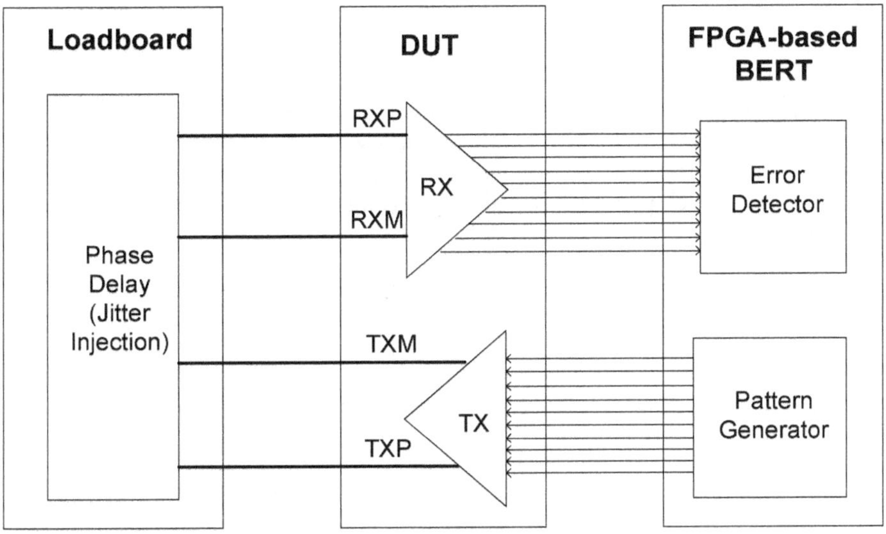

Fig. 5-6. Loopback-based jitter testing

In the above loopback testing scheme, we use the FPGA-based BERT discussed previously. The Pattern Generator in the FPGA provides parallel data to the transmitter. The Error Detector compares the recovered data with the transmitted data and records errors. The FPGA may also need to provide some glue logic. The FPGA-based BERT is not needed if we can still detect bit errors using the digital channels on ATE [2] or DFT features if they are available.

5.3.2 Phase Delay Based jitter Injection

A delay line or a phase delay line is a component where the input signal reaches the output of the component after a known period of time has elapsed [112]. The elapsed time or delay time ranges from femtoseconds to microseconds. Delay lines or phase delay lines have been widely used in electronics and derivative fields such as telecommunications and testing [97], [133].

Early delay lines were implemented with a RC-based ramp generator and a comparator that transitioned the delay line output when the ramp generator reached a certain voltage level. Calibration was done at the factory by blowing a serial fuses until the desired delay was achieved. The RC-based delay lines normally did not have temperature compensation provision. More sophisticated delay lines then were developed using a VCDL in conjunction with a compensation circuit to reduce delay variation across process, voltage and temperature [113].

Today, ultra-wideband phase delay lines have been developed using III-V technologies, such as InGap or InP Heterostructure Bipolar Transistor (HBT) devices [116].

InGap HBT is a proven reliable technology that has been widely used in large volume wireless applications. It exhibits characteristics such as high cutoff frequency, high linearity and temperature stability, suitable for ultra wideband device design. InP HBT has the highest cutoff frequency among the III-V available technologies [114], [115]. One unique product on the market is the phase delay line iT4036. It was developed by GigOptix [39] using high speed HBT Emitter Coupled Logic (ECL) topology realized in InP [116].

The iT4036 device is ultra-wideband, operating at speeds up to 12.5Gbps for data signals and 11.7GHz for clock signals. It can provide tunable phase delay up to 120-ps in a single device. Delay control can be either differential or single-ended and the delay control bandwidth is up to 1GHz. Its output amplitude is 400mVpp in single-ended mode and 800mVpp in differential mode. Figure 5-7 shows the device diagram of the iT4036 [116].

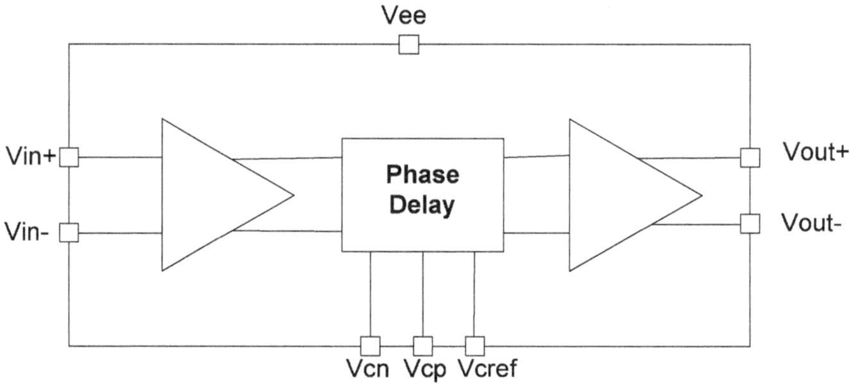

Fig. 5-7. Block diagram of iT4036

Traditionally, a phase delay line is only used to provide a constant delay controlled by a constant voltage in the delay control input. Considering the high bandwidth both in the signal path and the delay control path, we propose using the phase delay line for jitter injection to test HSSIs.

The main idea here is to apply an AC signal to the delay control pins. The delay for each data edge may be different, depending on the amplitude of the control signal at the instance of each edge. This is equivalent to injecting deterministic jitter to the input signal. Because the delay control bandwidth is up to 1GHz, we can inject DJ with frequencies up to a few hundreds MHz, suitable for HSSI jitter testing and characterization.

In our proposed jitter injection scheme, we connect the HSSI transmitter output to the input of the delay line. Then the output of the delay line is connected to the input of the HSSI receiver. Jitter is injected by connecting an AC control signal to its delay control pins Vcn and Vcp to dynamically control the phase delay between the input signal and output signal. By adjusting the amplitude of the delay control signal, we can control the DJ injected to the data signal.

Figure 5-8 shows the relationship between the delay control signal and the phase delay [116]. The relationship between the delay control voltage and the phase delay is very close to linear between the 20ps to 110ps delay range. Therefore, we can conveniently control the amount of injected DJ through adjusting the amplitude of the AC signal on the delay control pins in this linear range, and the following discussion refers to this range.

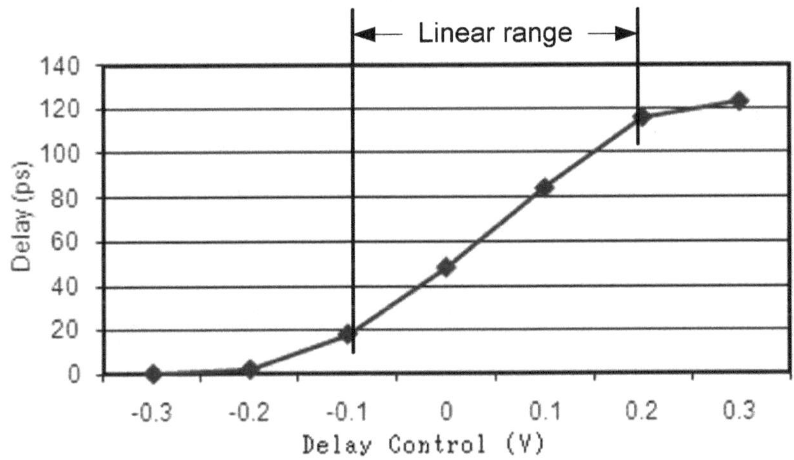

Fig. 5-8. Delay vs. delay control

To control the injected TJ in a linear manner, we need to primarily consider the RJ degradation issue. The RJ degradation is the additional RJ source introduced by the phase delay line. It can be straightforwardly characterized by obtaining the difference in RJ between the input signal and the output signal of the phase delay line. Figure 5-9 shows the RJ degradation for 12.5Gbps NRZ, 10.7Gbps NRZ and 12.5Gbps clock signals at different control voltage levels with Vcp tied to Vcref [116].

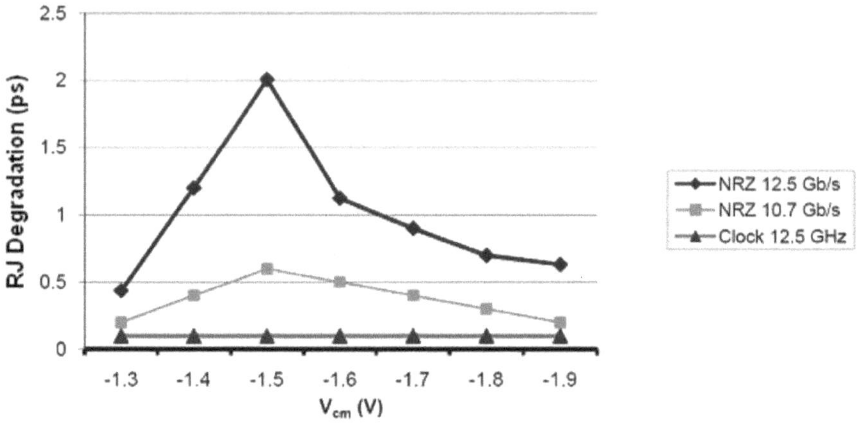

Fig. 5-9. RJ degradation vs. Vcm (courtesy of GigOptix)

Because the phase delay line shows a linear relationship between the injected DJ and the delay control magnitude, the RJ degradation needs to be constant in order to also keep a constant offset between the injected DJ and TJ. As discussed in Chapter 3.4, a constant offset between injected DJ (or PJ) and TJ, such as the one shown in Figure 3-17, enables us to translate TJ tolerance testing into DJ/PJ tolerance testing. To maintain a constant offset between the injected DJ and TJ, according to Figure 5-9, a clock signal is preferred when we use the phase delay based new jitter injection technique at the 10Gbps data rate range.

5.3.3 Experimental Results

We demonstrate the phase delay line-based jitter injection technique on an iT4036 evaluation board as shown in Figure 5-10. The input and output signals of the delay line on the evaluation board are routed to SMA connectors. In our experiments, we connect the transmitter output of the DUT to the input of the delay line through cables and then capture the output of the delay line using the digitizer available on ATE (the digitizer was discussed in Chapter 4). The delay control signal is provided by a digital channel on the ATE. Figure 5-11 shows the test setup. The digital channel provides a clock signal and the injected jitter can be adjusted by changing the V_{oh} and V_{ol} levels of the digital channel. The injected jitter frequency can also be adjusted according to test requirements.

Fig. 5-10. Phase delay line iT4036 evaluation board (courtesy of GigOptix)

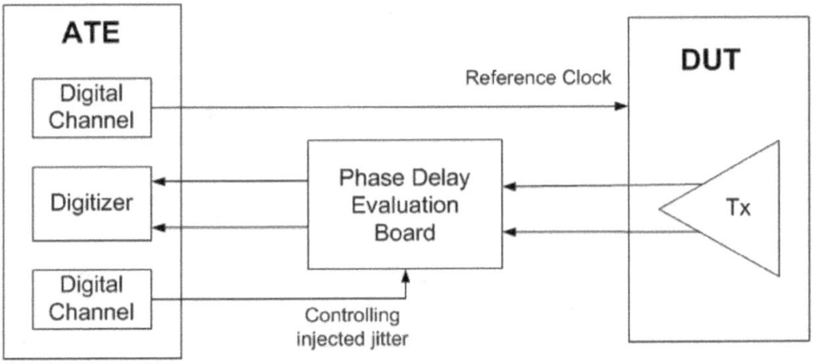

Fig. 5-11. Phase delay evaluation setup

Figure 5-12 plots the captured waveform of a 6 Gbps NRZ data signal from the output of the delay line with 400 samples for each bit. As we can see, the output signal is very clean. The rise/fall time is very short as specified in the delay line specification. The plot shows the output amplitude around 900mVpp, which is also close to the delay line specification (800mVpp).

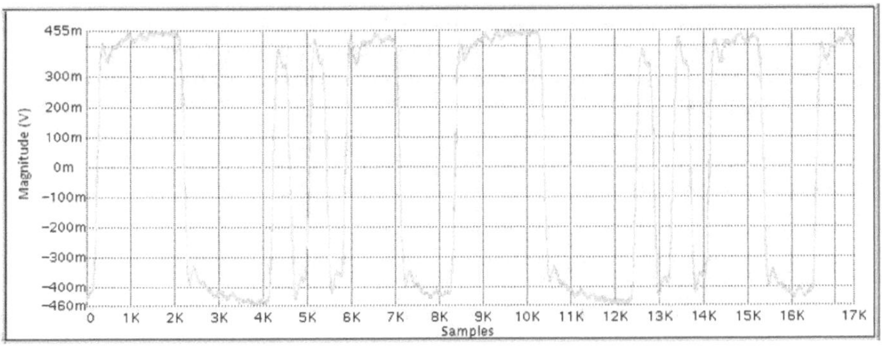

Fig. 5-12. The transmitter output waveform after the delay line

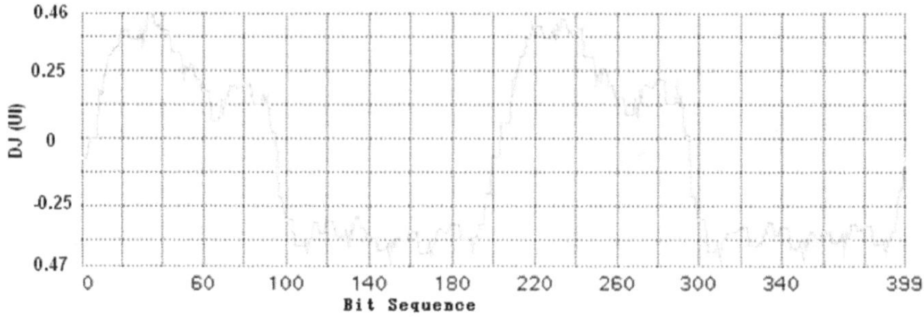

Fig. 5-13. Extracted DJ profile from captured data signal.

Figure 5-13 plots the extracted DJ profile from the captured data signal using the transmitter jitter extraction scheme proposed in Chapter 4.3. In the 400-bit 6Gbps data signal, the DJ profile repeats twice. Therefore, the DJ dominant frequency is 30MHz. As we can see from the captured waveform, the DJ profile is close to a square wave, which is the profile of the jitter source – a 30 MHz clock signal.

There is a minor distortion at the second half of the high logic level, which is likely caused by the intrinsic PJ of the device. The extracted DJ profile contains all DJ components, including the injected DJ and the device intrinsic DJ. As discussed in Chapter 4.3.4, the device also has an intrinsic PJ component at 30 MHz. This is why the DJ profile in Figure 5-13 is not exactly the same square waveform as we inject.

Figure 5-14 shows the extracted DJ values at different amplitude levels of the delay control signal. As we can see, we have an almost linear control of the injected DJ.

We can calibrate the DJ and TJ of the test signal at different delay control amplitudes using either bench equipment or the digitizer on ATE. Then we can use the same jitter tolerance extrapolation algorithm presented in Chapter 3.4 to accelerate the jitter tolerance testing.

The difference in the extrapolation process is that instead of varying the injected PJ to vary the TJ of the test signal discussed in Chapter 3, we vary the magnitude of the delay control to vary the TJ of the test signal in the proposed loopback testing scheme.

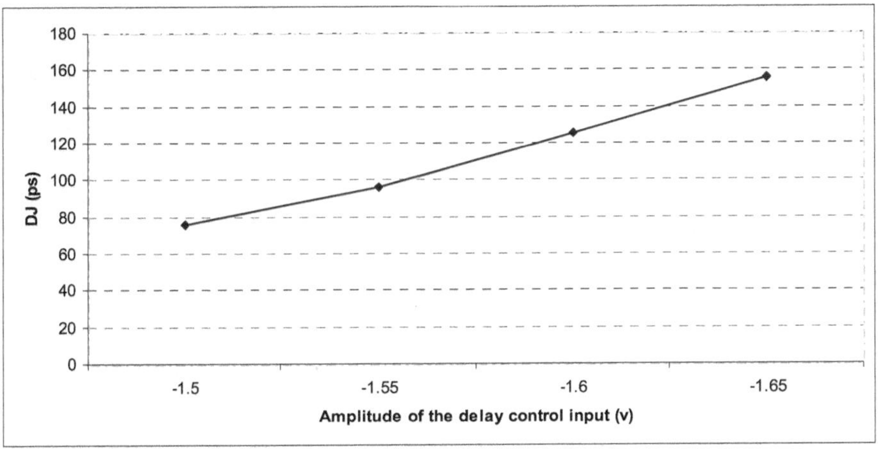

Fig. 5-14. Extracted DJ from the phase delay output

According to Figure 5-9, the clock pattern is preferred in the test setup because it has a constant RJ degradation over a wide range of delay control amplitudes and hence can keep a constant offset between the injected DJ and TJ. However, NRZ data signals can still be used in some applications, such as pass/fail testing, as we still can control the amount of injected TJ. Because the control is not linear for NRZ signals above 10Gbps, we cannot use NRZ signals for jitter tolerance characterization, where we need to report a jitter tolerance number as discussed in Chapter 3.4.2.

5.4 A Versatile HSSI Testing Scheme

In Chapter 3 and Chapter 4, we introduce ATE based HSSI test approaches. To overcome some limitations of the ATE-based solutions, we present an external loopback based testing solution using a new jitter injection technique in Chapter 5.3.

By utilizing high-speed relays, we combine all the solutions in Chapters 3, 4 and 5.3, and propose a versatile scheme in this section. The scheme provides more functionality and flexibility in post-silicon validation, characterization, testing and debugging of HSSIs. Figure 5-15 shows the block diagram of the proposed scheme.

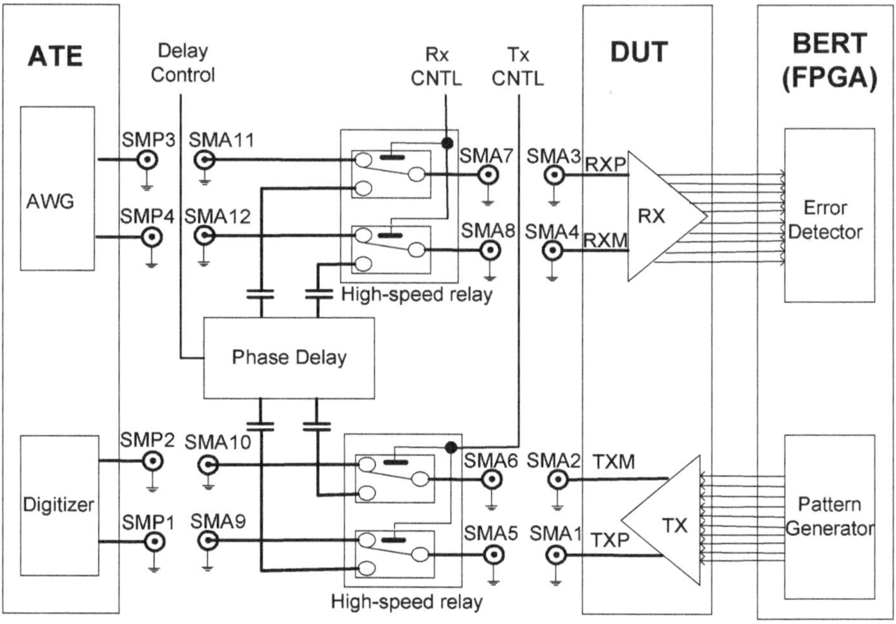

Fig. 5-15. A versatile scheme for HSSI validation and test – block diagram of the proposed infrastructure

5.4.1 Major Functions of our Setup

In Figure 5-15, the phase delay line and relays are incorporated on the testing loadboard. The relays are connected to either the phase delay line or SMA connectors. By changing the cable connections and controlling the relays, we can configure the proposed setup to realize different functions for testing, validation and debugging of HSSIs.

5.4.1.1 Testing, Validation and Debugging on ATE

In this configuration, the transmitter output is connected to the digitizer on the ATE, and the output of the AWG on the ATE is connected to the receiver input. Cables are used to:

- Connect SMA1 to SMP1, and SMA2 to SMP2
- Connect SMA3 to SMP3, and SMA4 to SMP4

This is the test setup that we mainly used tests undertaken in Chapters 3 and 4. Figure 5-16 plots this test configuration. The whole HSSI functionality and most design specifications can be qualified in less than one second in production [3], [4].

With this approach, we can accurately control the parameters in the receiver test signal, such as injected jitter, amplitude and test patterns. We can also capture the transmitter output waveform and extract transmitter parameters in a few tens of milliseconds.

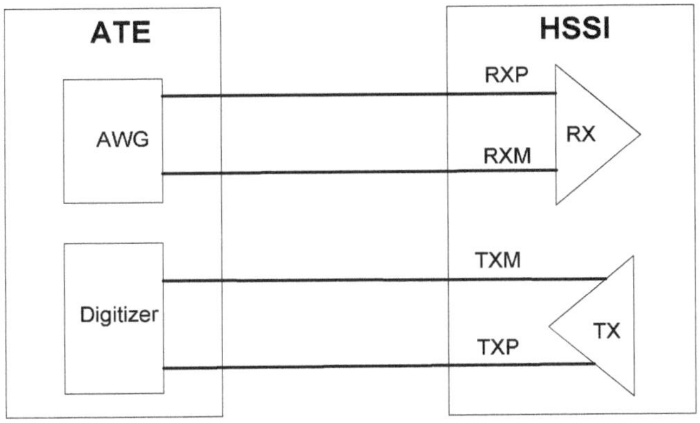

Fig. 5-16. Testing HSSIs on ATE

The ATE-based solution can also facilitate the HSSI validation and debugging process. For example, if we find receiver jitter tolerance is too low, we can quickly debug it by measuring the jitter tolerance at different conditions, such as varying the amplitude of the receiver input signal, changing the equalizer and PLL settings, turning on/off the transmitter block, sweeping supplying voltages, etc. All the measurement results are stored automatically and can be analyzed using ATE software. Using the ATE-based configuration, these kinds of debugging and validation procedures can be done more than 1000 times faster than traditional bench approaches [3].

5.4.1.2 External Loopback with Jitter Injection

This is the test setup shown in Figure 5-6 and discussed in Chapter 5.3. In the loopback configuration, Rx CNTL and Tx CNTL are set to low (the relays switch to lower throw). Delay Control can be connected to a digital channel on ATE or another resource to control injected jitter. Cables are used to:

- Connect SMA1 to SMA5, and SMA2 to SMA6
- Connect SMA3 to SMA7, and SMA4 to SMA8

5.4.1.3 Other Configurations

Few other configurations are possible using the proposed infrastructure, and we outline them here briefly.

1) *External loopback without jitter injection*: the DUT transmitter output is directly connected to the input of receiver. Cables are used to connect SMA1 to SMA3, and SMA2 to SMA4. This provides a quick way to check the functionality of the HSSI.

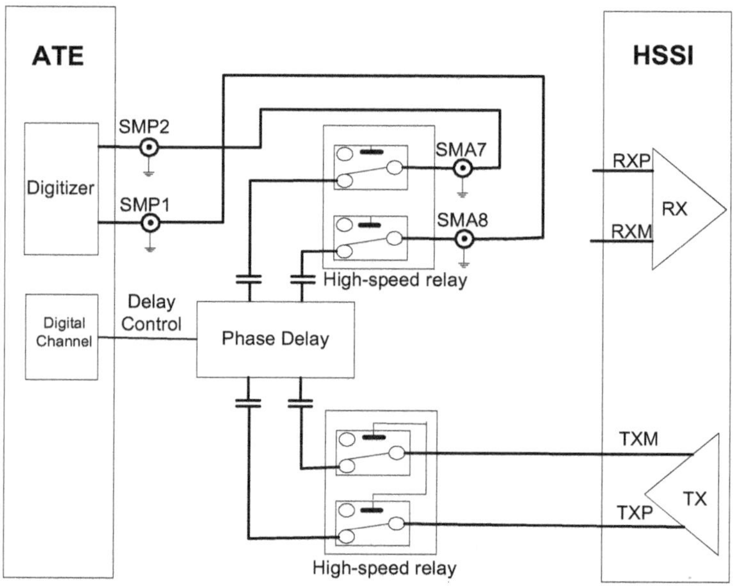

Fig. 5-17. Characterizing the relay and the phase delay using the digitizer

2) *Characterization of Relay and Delay line using the Digitizer.* The following lists the control settings and cable connections. Figure 5-17 plots the simplified connections of this configuration.

- Set Rx CNTL and Tx CNTL to low
- Connect SMA1 to SMA5, and SMA2 to SMA6
- Connect SMA7 to SMP2, and SMA8 to SMP1

3) *Characterization of relays only using the digitizer.* The following list the control settings and cable connections. Figure 5-18 plots the connections of this configuration.

- Set Tx CNTL to high
- Connect SMA1 to SMA5, and SMA2 to SMA6

- Connect SMA9 to SMP1, and SMA10 to SMP2

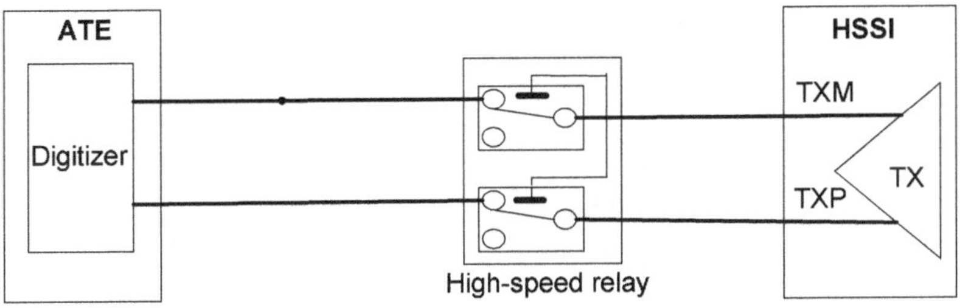

Fig. 5-18. Characterizing the relay using the digitizer

4) *Characterization of relays only or both the relay and the delay line using external instruments*: Instead of connecting to the Digitizer in 2) and 3), we can connect bench instruments to calibrate the relay and delay line.

In the above configurations, RF cables are used in order to maximize the configuration flexibility for validation, testing, debugging and calibration. This is very beneficial when we are in the debugging stage for a validation or test solution, such as evaluating the relay and the phase delay line. Once the solution is finalized, many cables can be replaced by PCB traces.

5.4.2 High Speed Relays

In the proposed versatile testing scheme, high-speed relays are used to switch between instruments (ATE or bench equipment) and loopback paths. When investigating signals with data rates at 6Gbps and above, it is challenging to maintain the signal integrity with relays inserted in the signal paths. One requirement for the relay is the bandwidth: it needs to be high enough to keep the signal characteristics after the signal passes through the relay.

There are a few kinds of relay technologies and each technology has its advantages and disadvantages in bandwidth, size, cost, reliability and life expectancy. A thorough survey of the relay technologies can be found in [117]. Considering we need to put the relays on the loadboard for high-speed applications, size and bandwidth are two major considerations. We concentrate our experiments on two kinds of relays: Micro-Electro-Mechanical System (MEMS) relays and electromechanic relays.

MEMS relays have received a lot of attention lately because of its high bandwidth and small size [118]. One disadvantage of the MEMS relays is that they are susceptible to damage from high in-rush currents or hot switching. When we started investigating the relay applications in multiple-Gigabit data applications, TeraVicta was developing a new MEMS relay TT1244. This relay offers unparallel RF performance in bandwidth (DC to 26.5GHz), insertion loss, and linearity.

Figure 5-19 illustrates the block diagram of the MEMS and Figure 5-20 shows its measured performance by TeraVicta [119].

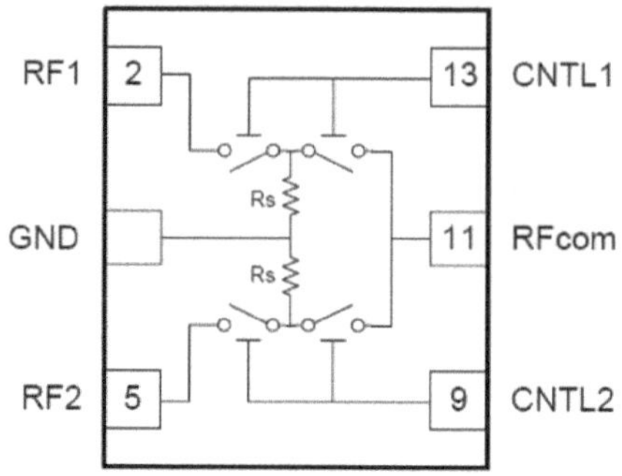

Fig. 5-19. TT1244 functional block diagram

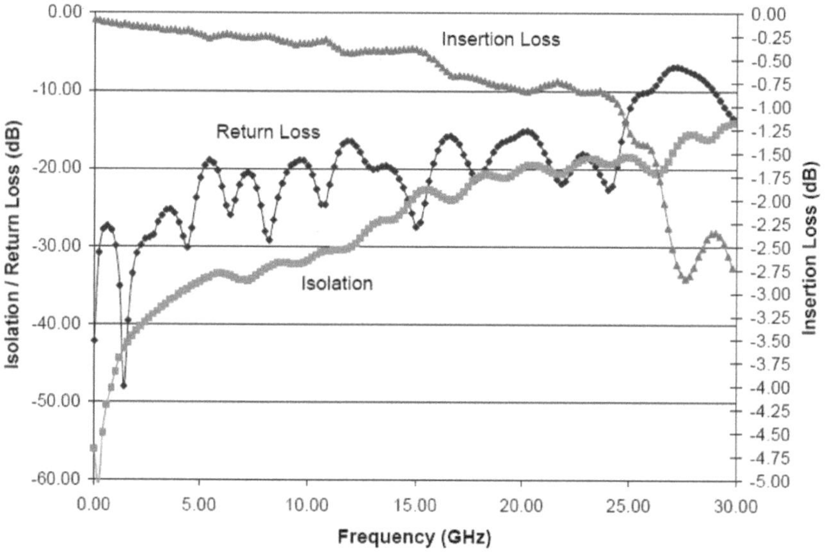

Fig. 5-20. TT1244 measured performance by TeraVicta

While the data rate of the HSSIs is continuing to increase and has reached above 10 Gbps, this switch provides the means for us to address signal integrity issues that are inherently very difficult to solve [120]. In addition, this switch has a very small footprint: it uses a standard surface-mount micro BGA chip scale package.

We therefore explored how TeraVicta's DC to 26.5 GHz switches can enable us to extend our test system's capabilities and keep pace with the increased frequency requirements of our new products. While TeraVicta was still developing the prototype of this switch, we developed our loadboard to include this switch and its control circuitry so we can evaluate the new product in a real ATE environment.

For MEMS switches, the control voltage is usually higher than that of other types of relays. The DC gate control voltage to CNTL1 and CNTL2 of TT1244 needs to be between 66v to 67v. Our ATE and loadboard do not have such a high voltage.

Therefore, we used a charge pump to generate the high control voltage. Figure 5-21 is the charge pump circuitry. The charge pump TT6820 only needs a power supply between 3v to 5v, which can be provided by any tester. The input signals (*In1~In6*) can come from any digital channels of the ATE or from an FPGA if no ATE instruments are available. The output signals (*CNTL1~CNTL6*) can directly control the MEMS relays.

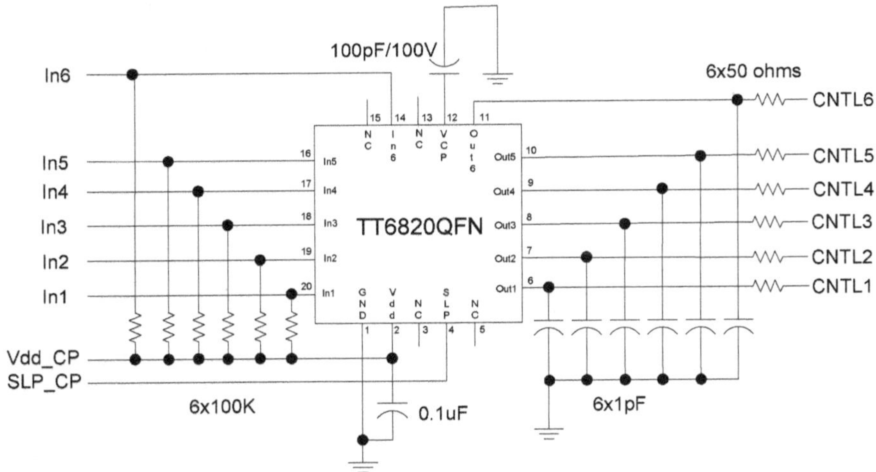

Fig. 5-21. Charge-pump circuit for the MEMS

When the TT1244 prototype was available, we were the first user evaluating the MEMS switch on a real loadboard. We put the switch in a 6Gbps signal path and compared the signal integrity of the path to the same length signal path without a switch. We did not observe any noticeable signal integrity degradation – the switch is nearly transparent.

However, we can easily notice some issues that hamper the usage of MEMS. The major one is the reliability. The MEMS relay worked well at the beginning, but it stopped working after a number of switching cycles. We carefully controlled our program to avoid the possible damage advised by the MEMS provider, but the MEMS relay still could get damaged. Even though the MEMS TT1244 product did not succeed, the performance it exhibited was very promising. MEMS would be a good direction to explore for applications with data rates at 10Gbps and above.

Currently, the data rate of the mainstream HSSI products is still below 10Gbps. Considering the issues in the MEMS switches and the current bandwidth requirements, we choose a hermetic electro-mechanic relay, Teledyne GRF300, for HSSI product testing after comparing performance, reliability and size [121]. Figure 5-22 shows the schematic and recommended layout [122]. The relay features a unique ground shield for each lead to ensure excellent isolation. The unique ground connection pushes the RF performance up into the 10Gbps data rates for signal integrity applications. In addition, its ultra-miniature size, high reliability and surface mount features make it a perfect choice for current HSSI testing applications.

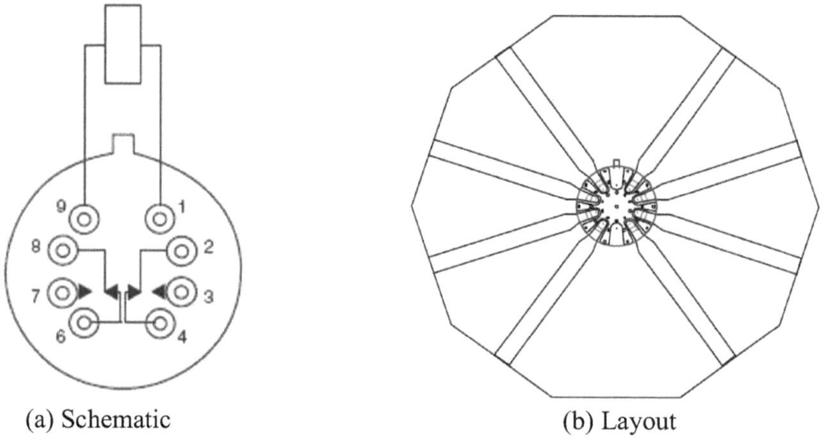

Fig. 5-22. GRF300 relay (Courtesy of Teledyne)

Figure 5-23 shows the typical signal integrity characteristics of GRF300 [121]. The eye diagram was captured by Teledyne using the Agilent AG86100 Digital Communication Analyzer. The data rate was set to 10Gbps and a 2^{31}-1 PRBS signal was used. The relay was mounted on an evaluation board. In the measurement, two 3-foot long RF cables were used to connect the evaluation board and the analyzer.

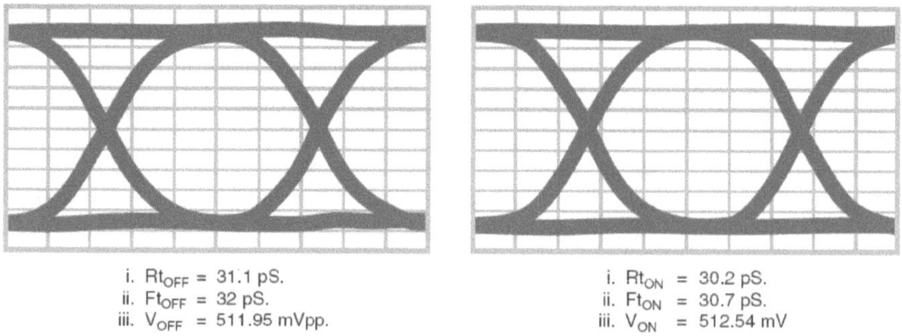

Fig. 5-23. Typical signal integrity performance at 10Gbps (courtesy of Teledyne)

We evaluated the signal integrity performance of the relay in our ATE environment. Figure 5-24 shows the evaluation setup. In this setup, the differential output signals of the transmitter are connected to the poles of the relay through RF cables.

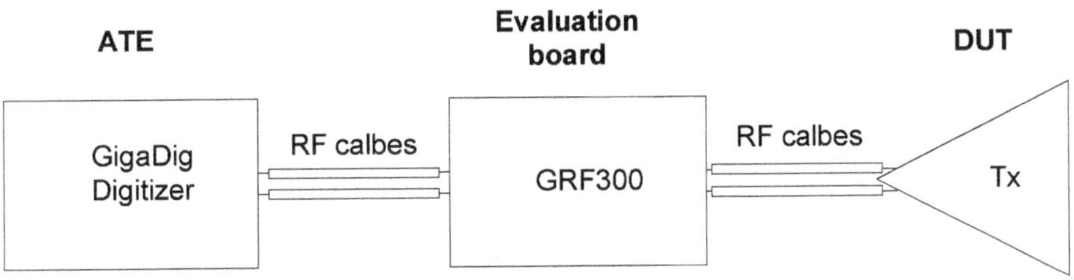

Fig. 5-24. GRF300 relay evaluation setup

The signals are then routed to a Gigabit digitizer on the ATE through the relay. Finally, the digitizer is used to capture the RF signal and extract the required signal integrity parameters, as we have already discussed in sufficient detail in Chapter 4.

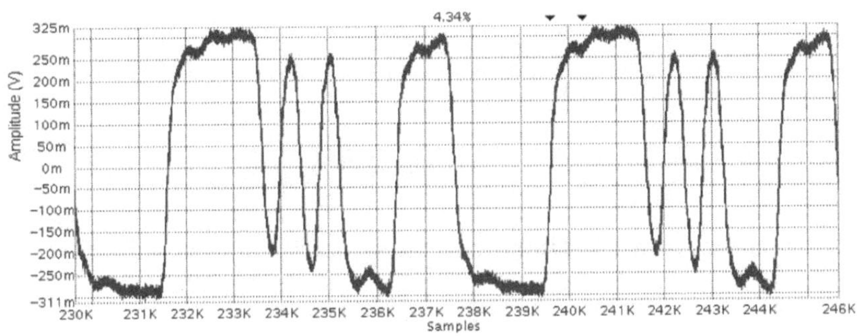

(a) Without relay in the signal path

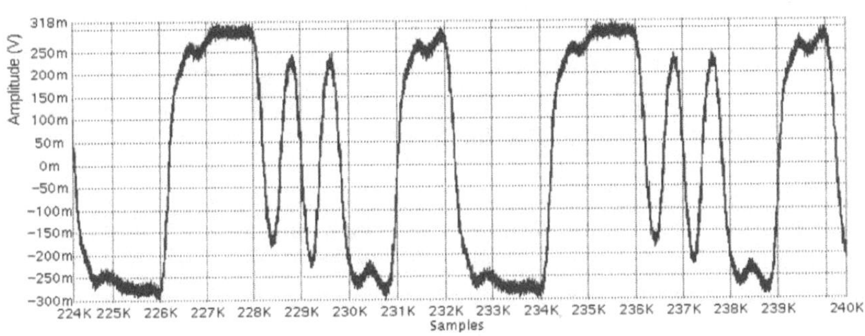

(b)With Relay inserted in the signal path

Fig. 5-25. Captured waveforms on ATE

In our setup, we set the data rate of our device to its maximum – 6Gbps. We capture the waveforms of the transmitter output after it passes through different signal paths. Figure 5-25 shows the captured waveforms with the relay inserted and bypassed respectively in the signal path. Table 5-1 lists the extracted signal parameters after the transmitter output passes through different signal paths.

As we can see, the signal still keeps similar characteristics after passing through an extra 50-inch cable and the relay compared to the signal passing through only an extra 20-inch cable.

The slight rising time increase is more likely caused by the extra length cable according to the difference between the original signal and the signal passing through an extra 20-inch cable.

Table 5-1. 6Gbps Signal Parameters in Different Signal Paths.

Signal Path /Parameters	Original	Extra 20-inch cable	Extra 50-inch cable + Relay
Amplitude (mV)	610	595	572
Rise (ps)	60	63.4	71.2
Fall (ps)	64	66.3	75.5
DJ (ps)	11.2	19.1	21.5
RJ (ps)	1.8	1.8	1.9

5.4.3 Limitations and Further Considerations

The proposed loopback testing scheme from this chapter removes the need for expensive high-speed ATE instruments, and is useful even when the ATE is being employed as well. The proposed infrastructure is especially suitable for multiple-line HSSI testing due to its low cost. Low cost testing has always been attractive the industry [79], [99], [137], and is a must for producing competitive, high-end products. Our scheme also overcomes the ATE instrument limitation for data rates above 6Gbps. However, there are also limitations and some special considerations need to be taken.

In the delay-line based loopback approach, one thing we need to consider is the amount of jitter that needs to be injected. In [3], we set the amount of injected jitter using a jitter extrapolation algorithm based on calibrated test signals. In the loopback approach, because the transmitter jitter may vary from device to device, the jitter in the receiver test signal is not constant anymore even with the same amount of injected jitter.

The loopback test cannot differentiate between the transmitter failure and receiver failure. To minimize the possibility of skipping bad devices or failing good devices, we need to characterize the transmitter jitter performance in order to set a proper amount of injected jitter. We can achieve a more accurate testing by using a gold device or by measuring the transmitter jitter and then setting the injected jitter accordingly. However, the test setup and the test program in both cases become complicated.

In addition, when implementing the loopback test in production, we also have to take the following into considerations:

- When multiple loadboards are used in a mass production environment, it is required to characterize the delay line to make sure that the characteristics variations of all the delay lines are within a permitted range.
- We need to calibrate the delay line on a regular basis to make sure its performance does not drift over time until the confidence to the new component is established.
- We need to characterize the delay line to get its delay control voltage vs. injected jitter relationship for each data pattern we plan to use.
- We still need to explore more sophisticated relays, such as MEMS for data rates above 10Gbps to minimize the signal integrity degradation.

For the FPGA-based bit error detection scheme discussed in Chapter 5.2, one drawback is that an FPGA device or board is needed if the current board does not have one.

However, it is important to notice that, in long term, the FPGAs can provide more testing functionality than just acting as a BERT. The BERT only takes a small portion of the FPGA resources [6]. We can then implement new testing algorithms or reduce the cost of the current testing solutions using the remaining FPGA resources. It is important to notice that the modern FPGAs are well capable of providing the necessary test and characterization circuitry, or even software run on "soft" processor cores at speeds that are sufficient for high-end circuitry under test.

6 BER Testing Under Noise

Abstract *To test BER under noise, we need a BERT and a noise generator. In further considerations, we will build on the BERT design already presented in Chapter 5, Section 5-1. This chapter introduces how to implement additive white Gaussian noise generators. Following the description of the underlying algorithms in Section 6-1, we propose a complete design in Section 6-2. As in some applications the distributions with long tail are a must, we show how one can achieve the excellent performance, both in the terms of the fidelity of the distribution to the Gaussian one, as well as the achievable tail length. We also present an application example for our noise generator.*

In the jitter tolerance testing, we stress the receiver using signals with controllable amounts of injected jitter. We accelerate the qualification of the jitter tolerance characteristics at 10^{-12} BER level by evaluating it at higher BER levels. Similar to the jitter tolerance testing, we can stress the system with controllable amounts of amplitude noise to test the BER performance under noise conditions. In jitter tolerance testing, one challenge is how to inject controllable amounts of jitter to the test signals. For BER testing under noise, the challenge becomes developing a noise generator with performance that meets test requirements. This chapter presents a new method of implementing AWGN generator/generators in FPGAs. The detailed implementation of the AWGN core, its performance and applications are discussed in this chapter.

6.1 AWGN Generation Overview

Existing methods of AWGN generation are based on a variety of statistical techniques. After reviewing existing methods and their drawbacks, we present our method.

Y. Fan, Z. Zilic, *Accelerating Test, Validation and Debug of High Speed Serial Interfaces*,
DOI 10.1007/978-90-481-9398-1_6,

6.1.1 Existing Methods

6.1.1.1 CLT Method

The CLT method is based on Central Limit Theorem (CLT). According to CLT, if X is a random real variable of mean m_x and standard deviation δ_x, the random variable X_N defined as

$$X_N = \frac{1}{\delta_x \sqrt{N}} \sum_{i=0}^{N-1} (x_i - m_x)$$

tends toward the Gaussian distribution of zero mean and the unity standard deviation, when N tends toward infinity. In the above expression, x_i, are N independent instances of the variable X.

Traditionally, the CLT method is implemented using an accumulator. The AWGN generator in [64] is based on this method. This generator consists of four M-sequence generators, three adders and an accumulator. The M-sequence generators are Linear Feedback Shift Registers (LFSRs) of lengths 28, 29, 30 and 31. By treating the last 10 bits of the shift register as a signed binary integer, a random number is generated. The AWGN generator produces one output every 12 system clock cycles by adding 48 10-bit random numbers. The output rate is 1 MHz.

If only the CLT method is used to generate Gaussian distribution, the convergence is very slow. Numerous independent random variables are needed to implement a high accuracy AWGN generator. In this case, either a very larger number of LFSRs and adders are needed or the output rate is very slow. So the CLT method is not suitable for high-speed applications.

6.1.1.2 Box-Muller Method

As a key tool in statistics, the Box-Muller algorithm can be applied to generate Gaussian distribution. This generator is shown in Algorithm 6-1.

1. Generate two independent random values x_1 and x_2, uniformly distributed over [0,1].
2. Obtain:

   $$f(x_1) = \sqrt{-\ln(x_1)}$$

   $$g(x_2) = \sqrt{2}\cos(2\pi x_2)$$

3. Generate Gaussian variable

   $$n = f(x_1)\, g(x_2)$$

Algorithm 6-1: Box-Muller Method

This method has the advantage of maintaining a one-to-one correspondence between the random numbers used and the Gaussian random variables produced, with every group of random values generated in step 1 producing one output in step 3 in Algorithm 6-1.

An FPGA implementation of the Box-Muller method is proposed in [47], where implementing *ln* and *cos* functions requires careful considerations regarding the number of recursions and relative position of points, as well as the precision of implementation. The efficient implementation is therefore not straightforward.

Another disadvantage of the Box-Muller method is that it is not suitable for generating high maximum output values. As indicated in Algorithm 6-1, the maximum output value of n is determined by $f(x)$, as $g(x)$ is bounded to the interval $[-\sqrt{2}, +\sqrt{2}]$.

Since f approaches infinity when the value of x_1 is close to zero, the maximum output value of n is determined by the smallest value of x_1. We express x_1 as 2^{-t}, where t is the number of bits used to represent x_1 (all bits represent the fractional part). When $t = 32$, the maximum output value of the generator is around 6.7; while t increases to 64, the maximum value can only increase to 9.4. Obviously, the hardware cost is high and the output speed is limited if we need to achieve good tail distribution using the Box-Muller method.

6.1.1.3 Mixed Method

A mixed method used to implement an AWGN generator in FPGAs is proposed in [47]. This method is based on the combination of the Box-Muller algorithm and Central Limit Theorem. The detailed hardware implementation and performance evaluation of the generator are presented in [47].

In terms of speed and accuracy, the proposed implementation is very efficient and has been adopted by industry to generate AWGN [60], [148]. However, the efficient implementation of the Box-Muller method is not straightforward as discussed in Chapter 6.1.1.2. Moreover, the implementation of Central Limit Theorem in [47] slows down the output rate by a factor of N, where N is the number of iterations; therefore, the mixed method decreases the AWGN output rate. For the generator proposed in [47], when N=4, the output rate is only 24.5 MHz, while its clock rate reaches 98 MHz.

6.1.1.4 Cellular Automata Based Method

The above existing methods all use LFSRs to produce pseudo-random numbers. LFSRs are very popular and effective for pseudo-random number generation, and have long been relied for generation of random numbers [149], [150]. However, when many sequences of random numbers are needed, the area consumed by LFSRs is large.

One good alternative is in using cellular automata to achieve a large variety of random number generators. In 1986, Wolfram [151] suggested that cellular automata could be used for efficient hardware implementation for random number generators. The generated random numbers can be transformed to Gaussian variables [152].

Cellular automata can be thought of as dynamic systems, discrete in both time and space [153]. The principle of cellular automata is that the next value of each register is calculated by a Boolean function from the current values of immediate neighbours and itself.

The Boolean transfer functions are referred to as the computation rules. Such rules have been categorized in a seminal work by Wolfram [153]. One of the setups that can generate m-sequences is a careful mix of Rule 90 and Rule 150 as shown bellow:

$$Rule-90: a_i(t+1) = a_{i-1}(t) \oplus a_{i+1}(t)$$
$$Rule-150: a_i(t+1) = a_{i-1}(t) \oplus a_i(t) \oplus a_{i+1}(t)$$

where $a_i(t)$ is the content of register i at time t. The positions of Rule-90 and Rule-150 in a register array can be determined according to [154], [155].

Due to its simplicity and regularity of design, cellular automata have been widely used for uniformly distributed random number generators [156], [157], [158], [159]. The transformation from uniform variables to Gaussian variables can be done based on CLT method. Another method for this transformation is illustrated in Figure 6-1 [152].

In Figure 6-1, an n-bit uniform variable is compared with the numbers in a Gaussian Cumulative Distribution Function (CDF) conversion table and then encoded to an l-bit Gaussian random number.

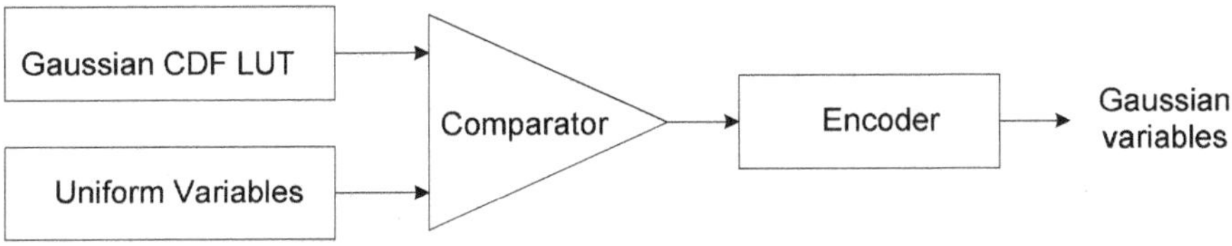

Fig. 6-1. Transformation from Random Variables to Gaussian Variables

This process is equivalent to grouping all of the points in the area under a Gaussian PDF to several columns, followed by randomly picking a point, and substituting the points with the one number that is the average value of the numbers in the column.

However, this transformation is usually difficult to implement for applications where high speed and high precision are required.

6.1.1.5 Analog Method

Even though the above digital implementations of AWGN generators have been successfully realized, converting these multi-bit digital signals to analog signals is challenging – it requires a high performance digital-to-analog converter.

An alternative is to utilize fully analog generators. Analog Gaussian noise generators are typically realized by low-pass filtering the output of a LFSR or by amplifying the thermal noise of a resistor. These kinds of generators are usually not programmable and not accurate for low BER testing.

In [66], a programmable analog Gaussian noise generator has been presented. The method encodes a specified Gaussian signal in a RAM, and filters the bit stream using an analog low-pass filter. The performance of the Gaussian noise generator realized in hardware is limited by the size of the available memory that needs to be initialized. Tradeoffs have to be made between the memory size and the signal quality. Typically the generator quality can only be guaranteed within 4δ.

6.1.2 Our Method

In order to overcome the disadvantages of the existing methods, we propose a novel method to implement AWGN generators. Our method consists of Polar method as shown in Algorithm 6-2 and our CLT method.

1. **Do**
2. Generate two independent random variables, U_1 and U_2, uniformly distributed over [0,1].
3. Set: V_1=(2*U_1)-1

   V_2=(2*U_2)-1
4. Set: $S = V_1^2 + V_2^2$
5. If S >=1, go back to line 2 and get new values for U_1 and U_2
6. **Loop** until $S < 1$
7. Set: $W = \sqrt{-2\ln(s)/s}$
8. Generate two independent Gaussian variables

   $X_1 = V_1 * W$

   $X_2 = V_2 * W$

Algorithm 6-2: Polar Method

As an improvement to the Box-Muller algorithm, Polar algorithm eliminates the trigonometric calculations. Polar algorithm provides a method to generate two independently distributed Gaussian variables with zero mean and the unity standard deviation [65].

For single channel emulation, we only need to generate one Gaussian variable (X_1 or X_2), but there are also the applications where both outputs can be utilized productively, such as in wireless signal modulations. The proof of the validity of this method is elaborated in [65].

Polar algorithm is faster than the Box-Muller algorithm because it uses few transcendental functions, even though it throws away, on average, 21% of numbers generated in the Do loop.

Our CLT method adopts the pipelined architecture instead of an accumulator usually adopted by the traditional CLT method. Therefore, our CLT method effectively eliminates the speed penalty while improving the accuracy of the AWGN generator.

6.2 Our Implementation

6.2.1 Generating Random Variables

According to Algorithm 6-2, the first step to generate a Gaussian variable is to generate two independent random variables, U_1 and U_2, uniformly distributed over [0,1]. In the past, the random variable generation was mostly done by software.

The software-based methods are well understood [160], [161], [162], but they frequently require complex arithmetic operations and thus are not feasible to be constructed in hardware. In this section, some techniques suitable for random number generation in hardware are first discussed, then the method used to generate U_1 and U_2 is introduced.

6.2.1.1 One Bit Random Number Generator

Ideally, the generated random variables should be uncorrelated and satisfy any statistical test for randomness. True randomness can be derived from certain physical phenomena, such as thermal noise in electronic circuit because of its well-qualified spectral and statistical properties. Figure 6-2 shows a representative implementation of a 1-bit true random variable generator [163]. In this circuit, the source *Vnoise*, which is the thermal noise of a precision resistor, is amplified and then passed to a high-speed comparator. The reference voltage of the comparator, *Vref,* corresponds to the mean voltage of the amplified noise signal. The output of the comparator is sampled and latched to a register. The latched 1-bit signal exhibits true randomness.

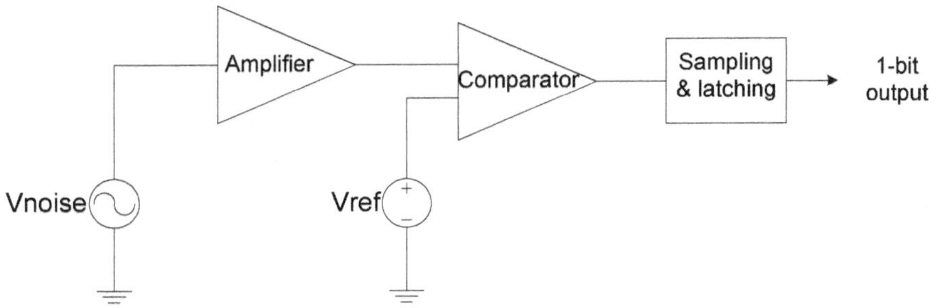

Fig. 6-2. A True 1-bit Random Variable Generator

The true random variable generator consists of mainly analog components and cannot be implemented by pure digital circuitry. The mixed-signal implementation significantly increases the system complexity and is relatively slow, so this method is not suitable for high-speed digital circuit design.

One common solution is to use linear feedback shift registers [164] to generate pseudo random variables. The sequence of a LFSR is based on specific relation between the feedback and feedforward values in the register. Though the generated pattern is actually repetitive and predictable, the sequence appears to be random if the cycle period of the LFSR is very large.

An LFSR uses feedback from the various stages of an m-bit shift register, connected to the first stage by means of XOR gates. The LFSR generating a single bit random number is based on the recurrence equation:

$$x_n = a_1 \cdot x_{n-1} \oplus a_2 \cdot x_{n-2} \oplus \cdots \oplus a_m \cdot x_{n-m}$$

Here, x_i is the ith number generated, a_i is a pre-determined constant that can be set to either 0 or 1, $\cdot$ is the AND operator, and $\oplus$ is the XOR (exclusive-OR) operator.

This relation hence implies that a new number (x_n) can be obtained by utilizing m previous values $(x_{n-1}, x_{n-2}, \cdots, x_{n-m})$ through a sequence of AND-XOR operations.

In an LFSR, the maximum achievable period is determined by m, which is 2^m - 1. In order to achieve the maximum period, a special set of a_is has to be used. In these sets, most a_is are 0; only two to four of them are 1. Thus, the actual recurrence equation is fairly simple, and the recurrence equations are different for different values of m.

Many references, such as [164], [165], have tables that list the recurrence equations exhaustively. Table 6-1 lists the recurrence equations for m with values from 2 to 8.

Table 6-1. Sample Recurrence Equations

M	Recurrence equation
2	$x_{n-1}\ x_{n-2}$
3	$x_{n-1}\ x_{n-3}$
4	$x_{n-1}\ x_{n-4}$
5	$x_{n-2}\ x_{n-5}$

6	$x_{n-1}\ x_{n-6}$
7	$x_{n-1}\ x_{n-7}$
8	$x_{n-2}\ x_{n-3}\ x_{n-4}\ x_{n-8}$

As an example of the recurrence equation implementation in hardware, the circuit of an LFSR with $m = 4$ is shown in part (a) of Figure 6-3. A four-bit shift register, with the signals from the first and fourth stages fed back through an XOR gate, generates 15 different patterns during successive clock cycles. If the initial value of the shift register is set to $q_3\ q_2\ q_1\ q_0$=1000, then the output of each register and the generated 1-bit output can be determined. The results are shown in part (b) of Figure 6-3.

As can be seen from Figure 6-3, in an LFSR implementation, an initial seed is needed to set the initial condition of the registers. The seed can be any state except for the combination of all zeros, which causes the random sequence to be stuck at zero forever.

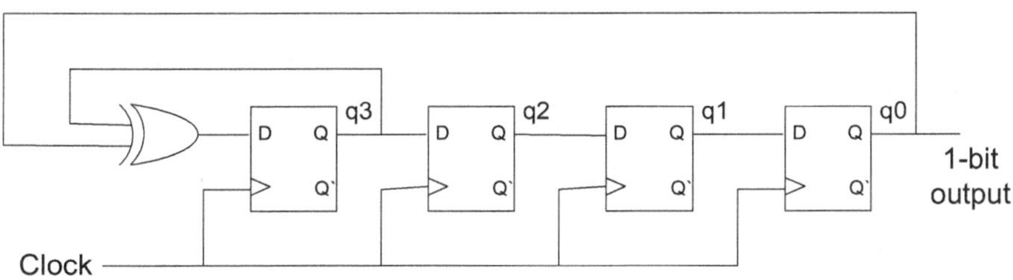

(a) Circuit

q_3	1	1	1	1	0	1	0	1	0	0	1	0	0	0	1	…
q_2	0	1	1	1	1	0	1	1	1	0	0	1	0	0	0	…
q_1	0	0	1	1	1	1	0	0	1	1	0	0	1	0	0	…
q_0	0	0	0	1	1	1	1	1	0	1	1	0	0	1	0	…
Output	0	0	0	1	1	1	1	1	0	1	1	0	0	1	0	…

(b) Generated Sequence

Fig. 6-3. LFSR-based Pseudo Random Number Generator

As can be seen from Table 6-1 and Figure 6-3, an LFSR-based random number generator only needs an m-bit shift register and one to three XOR gates, and thus the resulting circuit is very small in size, as well as its speed of operation is extremely high.

The generated sequence patterns have nevertheless the characteristics of randomly created numbers. Furthermore, since the period grows exponentially with the size of the registers, large non-repetitive sequences can be easily generated. For example, with a 64-bit generator running at 1 GHz, the period is more than 500 years.

6.2.1.2 Multiple-Bit Random Number Generator

It is also possible to generate multiple-bit random numbers using a LFSR. For example, one can use the LFSR in Figure 6-3 to generator 4-bit random variables (i.e. $q_3\, q_2\, q_1\, q_0$). However, the generated random variables are highly correlated and fail many statistical tests since a new random number keeps most bits from the old number and contains only 1-bit new information. To over the correlation problem, it is necessary to replace all bits in the random number rather than just one bit. One solution is to use parallel-LFSR method to generate multiple-bit random numbers. In this method, *m* independent LFSRs are used to generate *m*-bit random numbers.

Besides the parallel-LFSR method, there are other methods more efficiently utilizing FPGA resources to generate multiple-bit random numbers. For example, multiple-bit leap-forward LFSR method [166] is suitable for a small number of bits, and multiple-bit lagged Fibonnaci method [161], [162], [166], [167] is suitable for a large number of bits. However, their implementations are not as simple as the parallel-LFSR method. In addition, as discussed in Chapter 6.1.1.4, cellular automata can also be used to generate random numbers, and is especially suitable for generating a large number of random variables.

In our AWGN generator design, the parallel-LFSR method is used to generate random numbers U_1 and U_2. The FPGA resources taken by implementing U_1 and U_2 is very small. In the design, each of the two variables in line 2 of Algorithm 6-2 is set to be four bits in width, so four single bit random number generators are used to form a four-bit random generator. There are totally eight independent LFSRs used to generate the two 4-bit independent random variables (U_1 and U_2) in Algorithm 6-2.

The length of each of the LFSR is different, and all the LFSRs produce maximum periods. In this case, U_1 and U_2 are uniformly distributed between “0000” and “1111” (binary form). Please note that all these four bits represent the fractional part, so we get two independent random variables, U_1 and U_2, uniformly distributed over [0, 0.9375_D]. The maximum period of U_1 and U_2 is determined by the sum of all the lengths of the LFSRs, which can be adjusted to meet a required period.

6.2.2 Gaussian Variable Generation

In this section, the detailed implementation of AWGN generators is elaborated based on the generated random numbers U_1 and U_2. First the structure of a single AWGN generator is presented, then the structure of two AWGN generators is derived. Finally, a novel accuracy improvement method is introduced.

6.2.2.1 Implementing a Single Generator

Algorithm 6-2 shows that the Polar method can generate two independent Gaussian variables with a single iteration. It can also be simplified to fit the structure of a single AWGN generator. Figure 6-4 shows the block diagram of a single AWGN generator. In this implementation, pipelined structure is adopted to optimize the output speed.

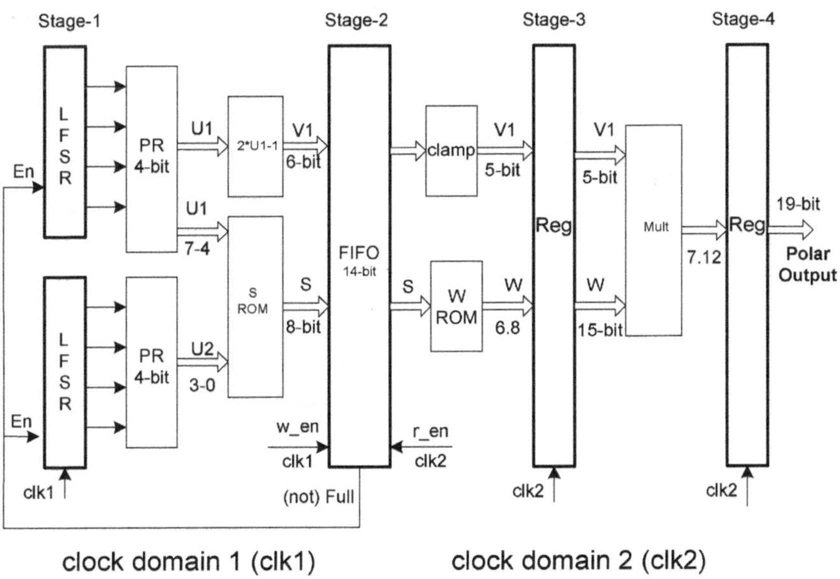

Fig. 6-4. Block Diagram of a Single AWGN Generator

6.2.2.1.1 Generating V_1 and S

The value for the variable V_1 is generated using signed adders performing the computation

$$V_1 = U_1 + U_1 - 1$$

Please note that before the addition, U_1 needs to be converted to a 6-bit signed number.

Computing S involves lots of additions and multiplications, which are very time-consuming. Most modern FPGAs include embedded RAM blocks. These blocks enable us to implement complex arithmetic operations with ROM-based designs, which are faster than the traditional arithmetic circuit implementations. Generating S takes advantage of this FPGA feature. ROM-based computation is used to implement the function

$$S = (2 * U_1 - 1)^2 + (2 * U_2 - 1)^2$$

The concatenation of U_1 and U_2 is set to be the address of the ROM and the values of S are set to be the data stored in the ROM. Both the address and data S are 8 bits in width. All the 8 bits for S represent its fractional part. If computed $S \geq 1$, the value of S stored in ROM is set to "00000000". As data "00000000" is only used to control the *w_en* signal of the FIFO (discussed in the next section), the effective range of data S is between "00000001" and "11111111" in binary form. In other words, the effective range of data S is over [0.00390625, 0.99609375] in decimal form.

6.2.2.1.2 FIFO Implementation

In Algorithm 6-2, a Do loop (line 1 to line 6) is used to generate qualified S, V_1 and V_2 for line 7 and line 8. On the average, line 1 to line 6 are executed 1.3 times of line 7 and line 8.

To achieve a constant output rate, a synchronizing FIFO is used. The job of the FIFO is to synchronize the implementation of the loop and the implementation of line 7 and line 8 in Algorithm 6-2 without losing or corrupting data. The width of the FIFO is 14 bits, 6 bits for V_1 and 8 bits for S. The loop implementation logic sends data to the FIFO receiver and the FIFO transmitter sends out data to the implementation logic of line 7 and line 8. The structure of the synchronizing FIFO is show in Figure 6-5.

The FIFO uses two clocks, *clk* for the receiver and *clk2* for the transmitter. When S is not equal to "00000000", *w_en* is enabled, the FIFO receiving V_1 and S at the rising edge of *clk*. Otherwise, no data is written to the FIFO and the next value of S is checked. In this case, the receiving data rate is a variable. In order to let the FIFO send data out at a constant rate (*clk2*), *clk2* must be smaller than the average rate of receiving data.

Hence, by setting the depth of the FIFO to be 16 and *clk2* to be half of *clk*, a constant output rate is achieved.

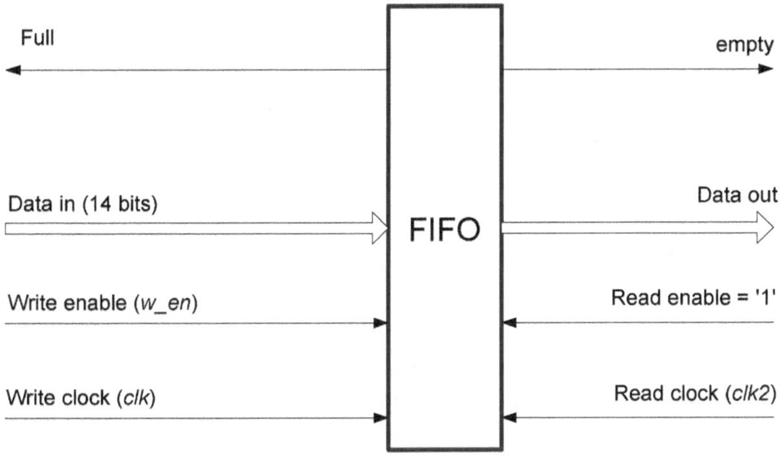

Fig. 6-5. Structure of the Synchronizing FIFO

In the FIFO design, four extra parameters (*read_pointer, write_pointer*, *counter* and *full*) are used to deal with the issues of synchronization, overflow and underflow. The parameter *counter* indicates how many locations have been filled with data in the FIFO according to the equation

$$counter = write\_pointer - read\_pointer$$

In situation of *write_pointer* = *read_pointer*, we do not know whether we have an empty FIFO or full one. To prevent this problem, we consider the FIFO full when 15 out of the 16 locations are occupied with unread data. When *counter* =0, it indicates the FIFO underflow. When *counter* = 15, it indicates the FIFO overflow.

The FIFO is guaranteed not to overflow by the following mechanism: when *counter* >= 14, *full* signal is asserted and the LFSRs are disabled. The FIFO *w_en* is disabled one clock cycle later.

In this case, the FIFO can still receive one group of data, so no data is missing. As the LFSRs are disabled, they stop generating data and no more data will be sent to the FIFO until *counter* < 14. Once *counter* < 14, LFSRs are enabled again and the FIFO *w_en* is enabled one clock cycle later if *S* does not equal to "00000000". By this way, the counter will always be smaller than 16. The FIFO will never really overflow.

With the above mechanism, the FIFO begins to send out data once *counter* reaches 14. On average, the possible rate of writing data to the FIFO is around 1.5 times faster than the rate of reading data from the FIFO when the clock rate of *clk2* is set to be half of the clock rate of *clk*. In this case, the FIFO, with a depth of

16, can still send data out even no data is written to it in 28 consecutive *clk* cycles. From the simulations, 5 was the maximum number of *clk* cycles in which no data was written to the FIFO (this number depends on the lengths and taps of LFSRs). In fact, a FIFO with depth of 8 is enough. We choose 16 to make our design more reliable. From the simulation results of 1 million clock cycles, the *counter* is always bigger than 10. It is concluded that the design is reliable enough to prevent underflow from happening.

6.2.2.1.3 Generating W

ROM-based design is also used to implement the function in line 7 in Algorithm 6-2

$$W = \sqrt{-2\ln(s)/s}$$

where the symbol *S* denotes the address line width of the ROM. The feasible width of *S* is 8 bits, and all the bits can represent the fractional part. The value of *W* is obtained from the data stored in the ROM. The plot of *W* as a function of *S* is shown in Figure 6-6.

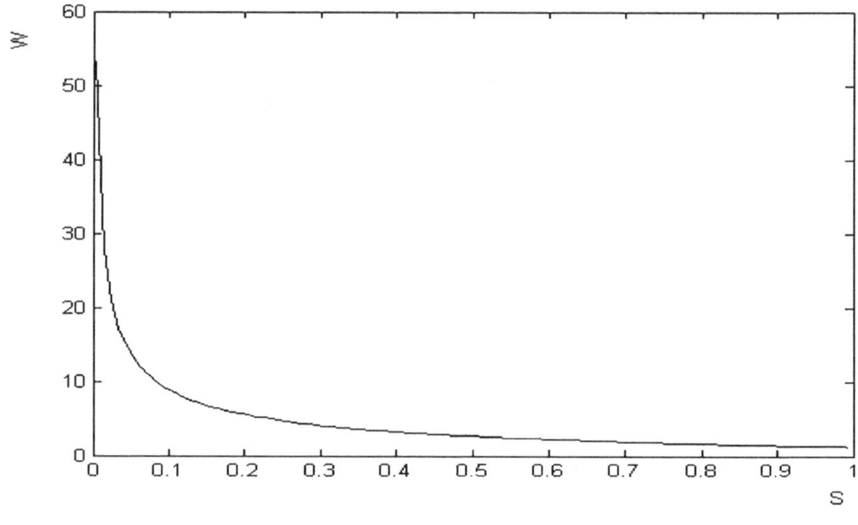

Fig. 6-6. Plot of Function *W*(*S*)

As *S* is between 0.00390625 and 0.99609375, according to Figure 6-6, *W* is between 54.2835 and 0.0886. To represent *W* in binary form, 6 bits are needed to represent its integer part. In our design, we use 14 bits to represent *W*, 6 bits for the integer part and 8 bits for the fractional part.

As the absolute value of V_1 from the FIFO is always smaller than 1, V_1 from the FIFO is clamped to 5 bits, 1 bit for the sign and 4 bits for the fractional part.

The register *reg3*, which stores the values of W and clamped V_1, is clocked by *clk2*, the clock rate of reading data from the FIFO.

6.2.2.1.4 Generating Outputs

The last step of implementing the AWGN generator is to implement the function

$$X_1 = V_1 * W$$

This step is completed by a single signed multiplier. Before performing multiplication, one *'0'* is concatenated to the most significant bit of W to convert W to signed form. The output of the multiplication is 19 bits in width, 1 bit for sign, 6 bits for the integer part and 12 bits for the fractional part. This output is sent to the output register *reg4*. The output of the *reg4* is what we need, which behaves like a Gaussian random variable.

The output of the AWGN generator can however be truncated to different widths, depending on application needs.

6.2.2.2 Implementing Two Generators

Figure 6-4 shows the structure of a single AWGN generator. For modulated data like QPSK signals, two noise generators might be needed for I and Q channels. According to Algorithm 6-2, the proposed one generator structure can be easily modified to implement two AWGN generators by adding V_2 implementation and another multiplier. The block diagram of two AWGN generators is shown in Figure 6-7.

In this structure, the width of the registers for each stage should be increased accordingly. As can be seen, the hardware cost is very small to add another AWGN generator based on the structure of a single generator. The proposed method of AWGN generation is especially suitable for multi-channel emulation.

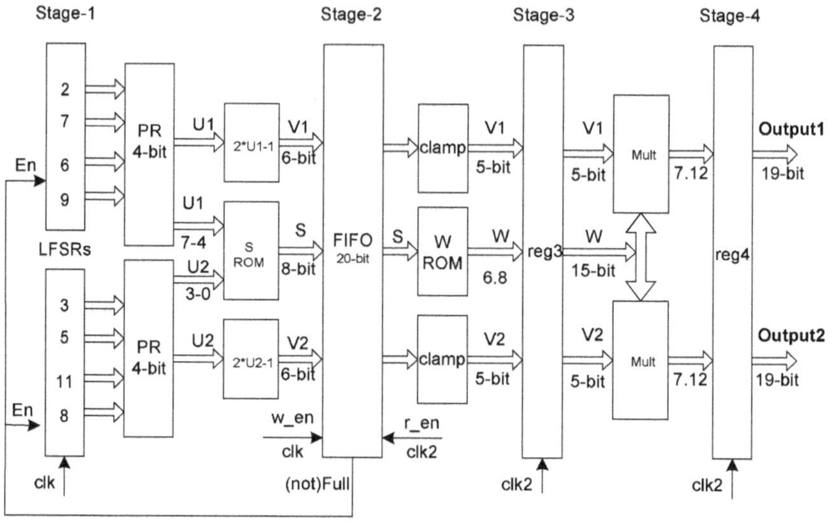

Fig. 6-7. Block Diagram of Two AWGN Generators

6.2.2.3 Accuracy Improvement

In our implementation, Central Limit Theorem method can also be used to smoothen the variation of the distribution when high accuracy is need. CLT method traditionally uses an accumulator. However, the accumulator will slow down the speed of the output. For example, when $N = 4$, where N is the number of random variables to be accumulated, the output rate after the accumulator is only one-fourth of that before the accumulator. As our implementation can produce two AWGN generators with little hardware cost, we can achieve one AWGN generator with better performance by simply adding the outputs from the two AWGN generators shown in Figure 6-7. This implementation does not incur the speed penalty.

To overcome the speed penalty problem, we propose a new CLT method for accuracy improvement. The block diagram of this method is shown in Figure 6-8, which implements the case when $N = 4$. The proposed scheme does not exhibit the speed penalty while improving accuracy.

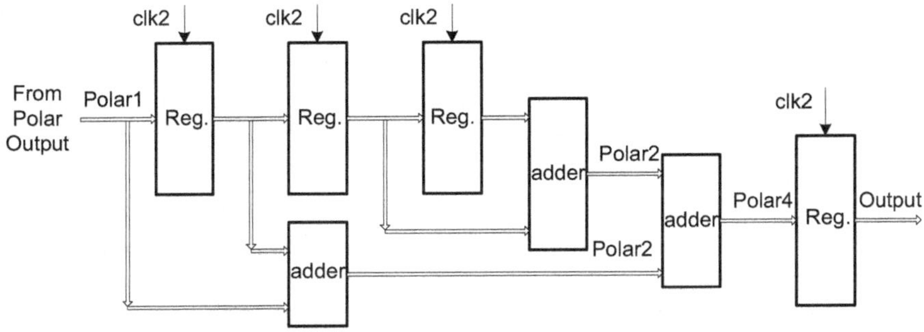

Fig. 6-8. New CLT Method

6.2.3 Statistical Properties of our AGWN Generator

There are a few methods that are used to analyze the statistical properties of a noise generator. The chi-square test and the Anderson-Darling (A-D) test are two goodness-of-fit tests adopted to evaluate some emulators [63], [123], [124]. For the chi-squire test, the axis is first quantized into segments. Then the actual number and expected number of samples appearing in each segment are determined. Based on the numbers of the samples, a single number is derived to serve as an overall quality metric of the generator. The chi-square test essentially compares the measured histogram to the theoretical histogram.

The A-D test concentrates more on the continuous time properties of a generator. It is a modification of Kolmogorov-Smirnov (K-S) test that gives more weight to the tails of the distribution than the K-S test does [125]. Other statistical methods include calculating $Q(x)$ and Kurtosis values, as will be elaborated in the subsection to follow.

Recall that the implementation of our AWGN generator is almost a direct map of the Polar algorithm without any partition or weighting [6], [67]. Because the Polar algorithm has been widely accepted for Gaussian variable generation in software, the statistical properties of our own hardware generator should be favorable as well. However, for proper evaluation and better insight, we explore the $Q(x)$ values and the *Kurtosis* value.

6.2.3.1 $Q(x)$ Evaluation

As the outputs of an AWGN generator are random variables with a mean of m_x and a standard deviation of δ , its performance evaluation should be based on statistics of the real outputs of the AWGN generator. Depending on the lengths of the

LFSRs used to generate U_1 and U_2, the period of the generator may reach the range of 2^n, where n is the sum of the lengths of the LFSRs. When n is equal to 50, the period is greater than 10^{15}. In this case, the complete verification of the AWGN generator should be based on the statistics of a very large number of samples, at least greater than 10^{15}.

Proper statistical evaluation the performance of such a larger number of samples needs a lot of hardware resources and time. Our experiments show that statistical results of thousands samples are a good approximation for the performance evaluation of the real AWGN generator. In the discussion to follow, we show the statistical properties of 10,000 and 500,000 samples from the output of our AWGN generator.

The process of getting the statistical properties consists of the following four steps:

1) Write the AWGN generator (VHDL top-level design) binary outputs to a text file.

2) The output data is imported to a *C* program that generates the probability density function of the outputs from the generator by sorting the outputs and computing the probability density $P[x_n]$ of each output.

3) The mean m_x and standard deviation δ of the AWGN generator are calculated from its PDF according to the following definitions.

Mean $$m_x = \sum_n x_n P[x_n]$$

Mean-Square $$m_x^{\;2} = E[x^2] = \sum_n x_n^{\;2} P[x_n]$$

Variance $$\delta^2 = E[(x - m_x)^2] = E[x^2] - m_x^{\;2}$$

4) $Q(x)$ of our AWGN generator is obtained according to

$$Q(x) = \sum_{i=1}^{n} x_i P[x_i]$$

where x_i, $i = 1, 2, \ldots, n$ are the possible discrete values from our AWGN generator that meet the condition of $x_i \geq x$; $P(x_i)$, $i = 1, 2, \ldots, n$ are the possibilities of x_i.

In this evaluation process, $Q(x)$ is the area under the tail of Gaussian PDF. It represents the probability that the Gaussian variable is between x and $+\infty$. The theoretical value of $Q(x)$ is computed according to

$$Q(x) = \frac{1}{\sqrt{2\pi}} \int_x^{\infty} e^{-t^2/2} dt$$

$$= \frac{1}{2} erfc(\frac{x}{\sqrt{2}})$$

where $erfc(x)$ is the complementary error function.

Table 6-2 shows the mean, variance and standard deviation of our generator. Displayed are the cases of 10,000 and 500,000 samples, respectively.

Table 6-2. Performance of our AWGN Generator

Samples	10,000	500,000
Mean	0.015835	0.008649
Variance	0.867620	0.866119
Standard Deviation	0.931461	0.930655

As we can see from Table 6-2, the relative error of mean and standard deviation of our AWGN generator is very small. For 500,000 samples, the relative error of mean is 0.008649 and the relative error of standard deviation is 0.069345. We can also see from the above table that the relative error of the mean decreases when the number of samples increases.

Table 6-3. $Q(x)$ Relative Error of our AWGN Generator

x	Theory Q(x)	Q(x) Relative Error of Our Generator		
		Figure 6-4	Figure 6-4 + Figure 6-8	
		10,000 samples	10,000 samples	500,000 samples
0	0.5000	2.76 %	1.02 %	0.24%
0.2	0.4207	-2.50 %	0.50 %	0.42%
0.4	0.3446	1.69 %	0.26 %	0.55%
0.6	0.2743	-0.10 %	1.06 %	0.80%
0.8	0.2119	1.88 %	1.74 %	1.09%
1.0	0.1587	4.70 %	3.21 %	1.20%
1.2	0.1151	-7.17 %	3.99 %	1.49%
1.4	0.0808	-4.29 %	4.95 %	1.85%
1.6	0.0548	-3.06 %	6.57 %	2.37%

The relative error of Q(x) of our AWGN generator is shown in Table 6-3. Relative errors are computed according to

$$relative \quad error = \frac{OurQ(x) - TheoryQ(x)}{TheoryQ(x)}$$

Table 6-3 shows the results of Polar algorithm only implementation (Figure 6-4) with 10,000 samples and the implementation combining Polar algorithm and Central Limit Theorem (Figure 6-4 + Figure 6-8) with 10,000 and 500,000 samples.

As can be seen from Table 6-3, our method with the parameters as shown in Figure 6-4 implements a high precision AWGN generator even with a limited number of samples. Our proposed CLT method shown in Figure 6-8 can further smoothen the variation of the distribution. We can also see from the above table that the relative error of Q(x) decreases when the number of samples increases.

6.2.3.2 Kurtosis Value

Kurtosis is also used to evaluate the Gaussian distribution [126], [127]. It is a measure of the flatness or peakedness of the probability distribution of a real-valued random variable. The *Kurtosis* value is often defined as the fourth standardized moment. For a sample of *n* values, the sample *Kurtosis* is

$$g_2 = \frac{m_4}{m_2^2}$$

$$= \frac{\frac{1}{n}\sum_{i=1}^{n}(x_i - m)^4}{\left(\frac{1}{n}\sum_{i=1}^{n}(x_i - m)^2\right)^2}$$

where m_4 is the fourth sample moment above mean *m*, m_2 is the second sample moment (variance), x_i is the ith value and *m* is the sample mean [128].

The *Kurtosis* value of an idea Normal distribution is 3. A *Kurtosis* value greater than 3 indicates that the distribution has a sharper peak and fatter tails, and the distribution is leptokurtic. If the *Kurtosis* value is less than 3, the distribution has a more rounded peak and thinner tails, and the distribution is platykurtic. Based on the 500,000 samples from the output of our AWGN generator, we calculated a Kurtosis value of 2.95, which is less than 2 percent off the theoretical value.

6.3 Baseband Transmission Testing

In this section, we present the baseband transmission testing setup and its test results in terms of BER as a function of SNR. The test setup mainly consists of the BERT core presented in Chapter 5-2 and the AWGN core presented in Chapter 6-2. The test results are very close to the theoretical values.

6.3.1 Baseband Signal Formats

In digital baseband transmission systems, there are various time domain signal formats. Figure 6-9 illustrates Return Zero (RZ), Non Return Zero (NRZ) and Non Return Zero Inverted (NRZI) signaling for the binary information data sequence 10011011. The NRZ and NRZI formats are commonly used in digital baseband transmission.

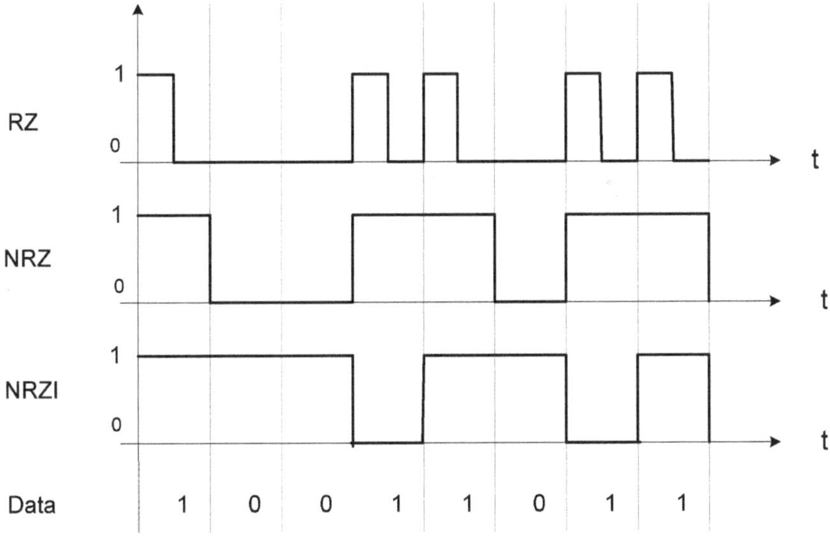

Fig. 6-9. Baseband Signal Formats

In a RZ transmission system, the binary information digit 1 is encoded as a high signal represented by 1, but the high signal returns to zero state before reaching the end of the bit interval. For illustrative purpose, it is assumed that the return occurs at the midpoint of the interval in Figure 6-9. In a RZ transmission system, the binary information digit 0 is encoded as a low signal represented by 0.

In NRZ format, the binary information digit 1 is encoded as a high signal represented by 1, and the binary information digit 0 is encoded as a low signal represented by 0. NRZ is the simplest baseband signal format. The NRZ modulation is memoryless and is equivalent to a binary Pulse Amplitude Modulation (PAM) or a binary PSK modulation in a carrier-modulated system [38]. The NRZ signaling format is more bandwidth efficient than RZ, as the pulses of NRZ signaling are wider than the RZ format.

However, there are two particular problems associated with NRZ transmission. First, when the transmitted data is static, which means there is no change from one bit interval to the next, there is no transition in the transmitted waveform. This causes timing problems when establishing bit synchronization. The second problem occurs with data inversion. If the levels of transmitted waveform are accidentally inverted during transmission, all the data is inverted, hence every bit is in error. Inversion can occur in several ways, such as a phase shift or losing track of the number of inversions.

To overcome these problems, NRZI format is introduced. NRZI signaling adopts differential techniques, in which the data is represented as changes in levels, rather than particular levels, of the signal. In NRZI format, the binary information digit 1 results in a signal transition, which can be either a low-to-high or a high-to-low; the binary information digital 0 results in no signal transition, which means the signal amplitude level remains unchanged. This type of signal encoding is called differential encoding. The coding operation is described mathematically by the relation

$$b_k = a_k \oplus b_{k-1}$$

where a_k is the binary information sequence into the NRZI encoder, b_k is the output of the encoder, and $\oplus$ denotes the exclusive-OR operation (addition modulo 2).

Based on the mathematical model of the processes involved, it is easily to get the structure of the NRZI encoder, as shown in Figure 6-10 (a), where Z^{-1} denotes one-cycle delay.

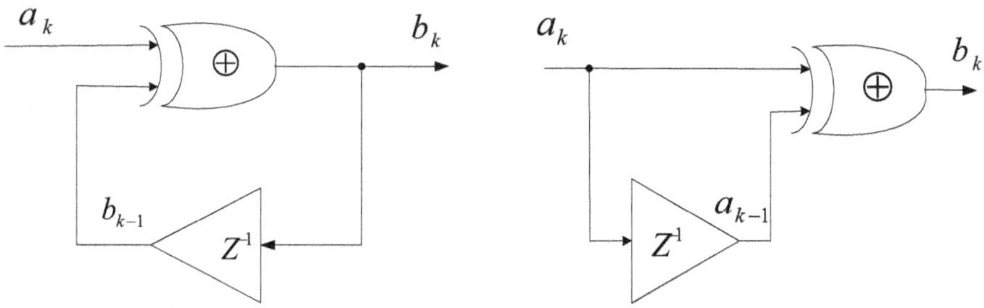

(a) Encoder (b) Decoder

Fig. 6-10. The Structure of a NRZI Encoder and Decoder

The NRZI decoder can be implemented as shown in Figure 6-10 (b). It compares the NRZI encoded signal to a delayed version of itself. If the two signals are the same in an interval, we know that 0 is being sent; if the two are different, a 1 is being sent. An exclusive-OR gate performs this decision process.

6.3.2 SNR Setting

Recall that from Chapter 2.3.1, when the correlation $\rho = 0$ in baseband transmission, we have $P_e = Q(\sqrt{\frac{E}{N_o}})$, where P_e is *BER*, and $\frac{E}{N_0}$ is *SNR*. In other words, the theoretical relationship between *BER* and *SNR* is characterized by

$$BER = Q(\sqrt{SNR})$$

As can be seen from the above equation, *BER* is only determined by *SNR* . In NRZ transmission systems, one (high level signal) is used to transmit data '1' and zero (low level signal) is used to transmit data '0'. We assume that data 1s and 0s have equal occurring probability, the average energy of two signals is

$$E = \frac{E_0 + E_1}{2} = 0.5$$

In an AWGN communication channel, the noise is Gaussian and characterized by a mean of zero and a variance of δ^2. The energy of the noise can be represented by

$$N_o = 2\delta^2$$

Though the above equations are derived from NRI signaling, they are also applicable to NRZI signaling. This is verified at the end of this case study. Combining the equations for E and N_o, we can get the equation for the *SNR* of baseband transmission. The *SNR* is determined by the variance of the noise and expressed as

$$SNR = \frac{1}{4\delta^2} \qquad (6\text{-}1)$$

In Chapter 6-2, an AWGN generator with zero mean ($m_x = 0$) and unity variance ($\delta^2 = 1$) has been developed. The variance of the generator can be changed to any value by adding a divider at the output of the generator. Suppose the original output is denoted by x, and it is divided by a, then the new variance of the generator becomes

$$\begin{aligned}\delta^2 &= E[(x'-m_x)^2] \\ &= E[(x/a)^2] \\ &= 1/a^2 \qquad (6\text{-}2)\end{aligned}$$

Combining equations 6-1 and 6-2, we have

$$SNR = \frac{a^2}{4}$$

where a is the scaling factor of the AWGN generator with zero mean and unity variance. By changing a, we can get different SNR conditions for the AWGN communication channel.

6.3.3 Testing Setup and Results

To test the BER under noise conditions, an AWGN generator with a variable gain is used to emulate an AWGN communication channel. The output of the variable gain AWGN generator is added to the output of the transmitter. The composed signal is then sent to the input of the receiver. We can use a BERT to generate test signals and compare recovered errors. Figure 6-11 shows the setup.

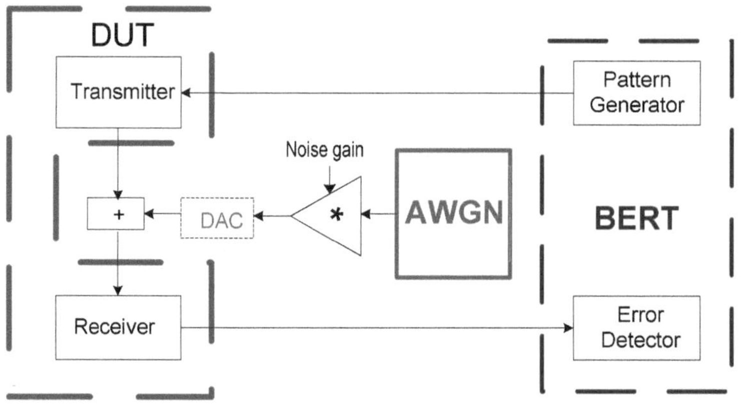

Fig. 6-11. Setup of testing BER under noise

In the test setup, the BERT can be a general purpose testing instrument that many companies provide [55], [56]. We can also use the FPGA-based BERT we present in Chapter 5.2. The DUT can be any communication interface or system that receives bit or word sequences and then restores the sequences after some signal processing or format changes. Some DUT examples include an HSSI and the integration of an encoder and a decoder.

The amplitude of the AWGN generator is programmable. Hence, we can emulate an AWGN channel in which signals are transmitted with different SNRs. The proposed setup can be used to test the BER performance of a real DUT in real operations under different SNR conditions. Because the output of our AWGN is digital, a Digital-to-Analog Converter (DAC) is needed if the AWGN output is added to an analog data signal. If the data signal is digital, the DAC is not needed. Figure 6-12 illustrates the BER setup for a NRZ digital baseband system, where no DAC is needed.

In the testing setup shown in Figure 6-12, the AWGN generator block and the BERT block are the IP cores introduced in Chapter 5-2 and Chapter 6-2, respectively. This testing setup also constitutes a digital communication system. In this system, the transmitter consists of the pattern generator; the communication channel consists of the AWGN generator, the divider and the adder; the receiver consists of the comparator and output decision block. The error detection block is used for performance evaluation.

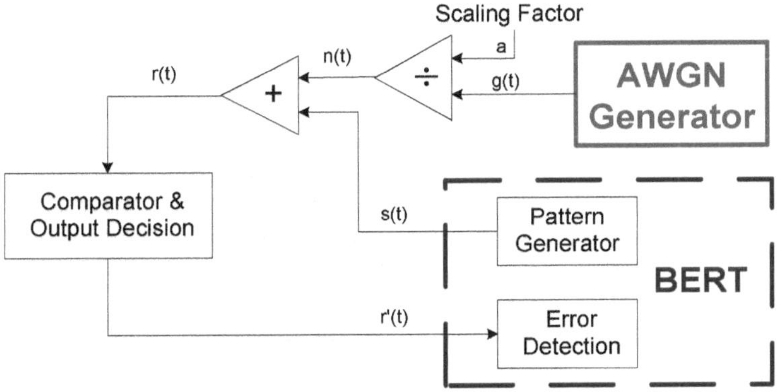

Fig. 6-12. BER Testing Setup for NRZ Digital Baseband

The output from the pattern generator is PRBSs. The output sequence, denoted by $s(t)$, is the data signal to be transmitted. The data signal $s(t)$ is corrupted by the noise signal $n(t)$ in the AWGN communication channel. The noise signal $n(t)$ is derived from $g(t)$ by dividing $g(t)$ by the scaling factor a, where a is six bits in width, five for integer and one for fraction. By setting the value a, the SNR condition of the communication system can be set. The noise signal $g(t)$ is the output of the AWGN generator with zero mean and unity variance. As discussed in Chapter 6-2, the noise signal $g(t)$ is 19 bits in width, 1 bit for the sign, 6 bits for the integer and 12 bits for the fraction.

The noise corrupted data signal $r(t)$ is compared with a threshold to determine the output of the receiver. In NRZ transmission system, 0s and 1s are transmitted and they have equal occurring probability; therefore, the threshold is set to be 0.5. If $r(t)$ is bigger than 0.5, $r`(t)$ is set to be 1; otherwise, $r`(t)$ is 0. Finally, the received bit sequence $r`(t)$ is compared with a delayed transmitted sequence $s(t)$ bit by bit, and errors are counted. The measured BER is obtained as the ratio of the counted errors and the number of transmitted bits.

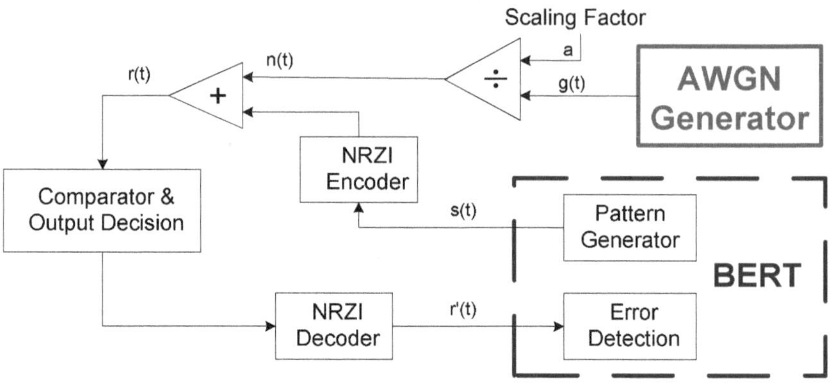

Fig. 6-13. BER Testing Setup for NRZI Digital Baseband

In the above digital baseband testing, the signal format is NRZ. For a NRZI baseband communication system, the testing setup shown in Figure 6-13 is used to test its BER performance. In this test setup, the structure of the NRZI encoder and the NRZI decoder is the same as these shown in Figure 6-10. The other blocks have already been introduced in the NRZ digital baseband testing setup.

Table 6-4 lists the BER test results. The measurements were taken while running the AWGN core and the BERT core in an Altera Mercury FPGA board. Measurements were also taken using an Agilent 81200 BERT to further verify our BERT.

Table 6-4. BER Measurements for Digital Baseband

a	SNR (dB)	Total bits	Our BER	Agilent BER	Error (%)
2	0.6	2024	1.62e-1	1.65e-1	1.85
3	4.1	12448	6.58e-2	6.61e-2	0.45
4	6.6	12448	2.32e-2	2.40e-2	3.33
5	8.5	20000	5.65e-3	5.32e-3	4.32
6	10.1	1000000	1.20e-3	1.31e-3	8.40
7	11.5	1000000	1.76e-4	1.83e-4	3.83
8	12.6	10000000	1.62e-5	1.78e-5	8.99

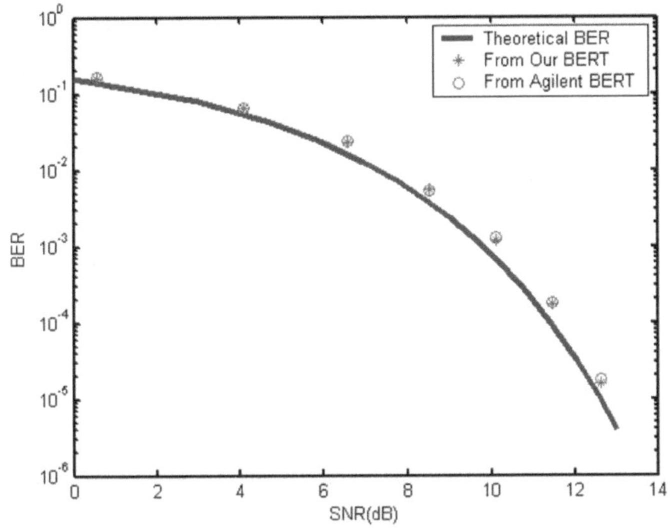

Fig. 6-14. Measured BER vs. theoretical BER

Figure 6-14 shows the plot of the measured BER and theoretical BER of digital baseband as a function of input SNR. Recall from Chapter 2.3.1, the theoretical BER of digital baseband is given by

$$BER_{theoretical} = Q(\sqrt{SNR}) = \frac{1}{2} erfc(\sqrt{\frac{SNR}{2}})$$

In Table 6-4, the following equations are used. These equations have been discussed previously.

$$\delta^2 = \frac{1}{a^2}$$

$$SNR = \frac{a^2}{4}$$

$$BER_{measured} = \frac{ErrorCout}{TransmittedBits}$$

In the testing setup, the pattern is generated using an LFSR, which produces more 1s than 0s, as all zeros is not a valid pattern while all ones is. Therefore, the actual signal energy is bigger than the theoretical value 0.5; for example, if the length of the LFSR is 3, the signal energy it produces is 4/7. We take this factor into consideration when calculating the SNRs in Table 6-4.

As indicated in Figure 6-14, the measured BER using our BERT perfectly coincides with that from the expensive Agilent BERT, and is close to the theoretical BER. The plot demonstrates the validity of the AWGN generator and the testing scheme

In the above testing, it takes less than one second to generate the point at 1.62e-5 BER, while software simulations take hours. Furthermore, regardless of the noise generation scheme, going down to low error rates requires many samples just to exhibit errors - for 10^{-12} BER at 1 Gbps data rate, it takes three hours (assuming running 10^{13} bits).

In production testing, the normal practice to qualify the BER performance at such low levels is through extrapolation [2]. However, if direct BER measurements at 10^{-12} or lower are needed, our AWGN is a good candidate as we will discuss in the next section.

Although the above experiment is based on testing a digital baseband system, the proposed BER testing scheme can be applied to other AWGN transmission system. Depending on the modulation scheme of the system, other components, such as a digital-to-analog convertor and an attenuator, might be needed. Furthermore, the AWGN module can be modified to emulate more complex channels, such as Rayleigh channels.

6.4 Advantages of Our AWGN Generator

For the AWGN generator we developed and discussed in Chapter 6.2, there are a few advantages compared to other implementations. The prominent advantage is its large maximum output value.

Theoretically, the tail of an AWGN distribution should extend towards infinity and the maximum output value is infinity. In reality, it is impossible to generate an infinity number. For an AWGN emulator, the tail is bounded by its maximum output value *m*.

Most existing standalone AWGN emulators are based on the Box-Muller Algorithm, through which is difficult to achieve a large maximum output value. Utilizing the Box-Muller method presented in [47], a commercial AWGN core has been developed [60]. This core is only capable of generating a maximum *m* value of 4.7.

Two years later, Lee et al. advanced the Box-Muller method [61] and increased the maximum *m* value to 6.7. This development has improved a bit the tail distribution of the AWGN generator. However, the price paid for this improvement is in quadrupled hardware resources, while the speed is halved. In 2005, this value was increased to 8.2 by using sophisticated FPGAs and implementation techniques [129].

The most recent advance in implementing AWGN generators in FPGAs based on the Box-Muller Algorithm is presented by Alimohammad *et al.* from University of Alberta in [63], where the tail distribution is extended to ±15 times of the standard deviation δ. The paper [63] also gives a comprehensive overview of Gaussian noise generation algorithms and related work on implementing the generators in hardware.

As a key advantage, the Polar method we developed can easily achieve high maximum output values with little hardware. As indicated in Algorithm 6-2, the maximum output value is calculated as a logarithm of the minimum value of the square operation of the random variables U_1 and U_2. In contrast, the Box-Muller method calculates a logarithm of the random variable x_1 directly as shown in Algorithm 6-1, which requires a tremendous amount of hardware to achieve a big output value.

In our implementation, we can produce a maximum output value of 53.3 by using only 4 bits (all of them to be used for the fractional bits) to represent each uniform random variable (U_1 and U_2 in Algorithm 6-2). In contrast, we note that to achieve such a high maximum output value using the Box-Muller method, we will need to use more than 1000 bits of information to represent the uniform random variable x_1 in Algorithm 6-1, which is almost impossible to implement in hardware.

A noise generator with a large maximum output value is essential for low BER evaluation. According to Equation (2-11), the BER and SNR are linked by Q factor:

$$BER = Q(\sqrt{SNR})$$

Therefore, low BERs require high SNRs. In an AWGN communication system, the SNR is determined by the standard deviation δ of the AWGN generator when the output of the generator is added to a data signal with constant energy. In order to emulate a low BER system, we usually scale down the distribution of the Gaussian generator to achieve a low standard deviation.

However, some samples of the noise generator output must be large enough to produce bit errors. The lowest BER that an AWGN generator can emulate is determined by the tail distribution of the AWGN generator, or more specifically limited by the maximum output value m of the noise variable. For a digital system where data "0" and data "1" are transmitted, the maximum output of the noise generator is required to be bigger than the threshold (usually 0.5) in order to generate bit errors.

Most existing hardware AWGN generators (e.g., [47], [60], [61] and [62]) only have a maximum output value of 7. If we use such generators in baseband transmission evaluation, they can only generate a theoretical maximum SNR of 16.9dB by being scaled by a factor of 14 (the noise scale factor a) according to Equation 5-2.

The 16.9dB SNR can be translated into a BER around 10^{-12} according to Equation (2-11), but such generators are only suitable for exploring channel behavior at BERs down to the range of 10^{-9} to 10^{-10} [61]. For any AWGN generator with bounded output, the distribution near its maximum output value is not Gaussian anymore and cannot be used because ideal Gaussian distribution is unbounded (the maximum output value is infinity). Hence, the tail distribution of these AWGN generators needs to be improved for applications with low BERs, such as 10^{-12}.

The latest AWGN generator presented in [63] can generate a maximum output value of 15. The generator is capable of evaluating BERs below 10^{-12}. Our AWGN generator further increases the maximum output value to 53.3. Therefore, our AWGN generator is a good candidate for very low BER applications, or alternatively, for distributions with potentially long tail.

Besides the maximum output value, another advantage of our AWGN generators is that it is highly scalable. In the implementation discussed in Chapter 6.2, we only use 4 bits to represent the uniformly distributed random variables, which is a remarkable feature.

The lengths of the LFSRs used to generate each bit are also short. They can be easily increased to achieve better performance if needed. In addition, our noise generator is relatively straightforward to implement.

One challenge of the Polar method implementation is to convert the variable output rate to a constant output rate. We resolve it by inserting a dual clock FIFO. Other function implementation is almost a direct mapping of Algorithm 6-2. Due to the bounds on the intermediate values, all of which lie within the unit value interval, it can be easily shown that no further partition, rounding or approximation is needed in the execution of the algorithm.

7 Conclusions

Abstract *We summarize here the contents of the whole book and provides a perspective on the practicality and usefulness of the obtained high-speed serial interface test and characterization results.*

This book has presented schemes that can accelerate the qualifications of jitter characteristics and BER performance of HSSIs. To deal with both the HSSI transmitter and the receiver side test and characterization, we have developed systematic ATE solutions for the characterization and compliance qualification of receiver jitter tolerance and transmitter output jitter.

For the receiver jitter tolerance testing, we have proposes a jitter tolerance extrapolation algorithm, capable of accelerating jitter tolerance testing by a factor of thousand. This improvement makes it possible to run the time-consuming jitter tolerance test in production testing of massively produced microelectronic devices. The test signal for such tests is generated by the AWG on ATE. The BER can be monitored either by the digital channels on ATE or by the BIST circuitry. This approach does not need any add-on hardware modules. Based on the AWG6000, we conduct jitter tolerance testing investigation on 1.5Gbps and 3Gbps signals, primarily using SATA devices. Higher speed applications are attainable using a higher performance AWG.

Further, for transmitter jitter testing, we have proposed three realistic approaches to extract the jitter components. With the proposed schemes, we can manage to have the whole transmitter test done within 100ms and achieve a jitter accuracy within +-0.5ps (in terms of run-to-run variation, and the difference between bench and ATE measurement results), which no one else has ever achieved, to the best of our knowledge.

Extensive experiments have been done on 3Gbps and 6Gbps signals to verify the accuracy and test repeatability. Based on a high-bandwidth digitizer available on ATE, this solution too does not need any add-on circuitry. The same methodology can be applied to test any HSSI transmitter.

Even though the book only addresses the transmitter jitter testing and receiver jitter tolerance testing, other HSSI tests are either simply by-products of the two tests or are very straightforward to conduct.

For most other transmitter related tests, they can directly be done based on the captured transmitter waveform. For example, the transmitter function can be veri-

Y. Fan, Z. Zilic, *Accelerating Test, Validation and Debug of High Speed Serial Interfaces*,
DOI 10.1007/978-90-481-9398-1_7,

fied by comparing the captured values to expected values; the transmitter level can be obtained by checking the captured level difference between 00000 and 11111, and the rise/fall time can be obtained from any edge with a full swing (in low transition density areas). In addition, we can easily add more captured samples to test other transmitter functions/parameters, such as transmitter Output Enabled (OE), common mode level and pre-emphasis.

The receiver jitter tolerance test actually also covers significantly the receiver functionality testing. If the receiver does not function, we could not get any jitter tolerance data.

Other receiver parameter tests can be done using the same test setup with some minor adjustments. For example, for the receiver tracking range test, we only need to adjust the sampling frequency of the AWG to source a signal with a frequency offset that we need; for the Out of Band (OOB) test, we only need to change the AWG pattern and amplitude for different turn-on/turn-off level settings.

The whole transmitter and receiver testing solution can be used to test any HSSI devices. It is independent of the detailed structure of the device to be tested, so in that sense can be considered to be high-level design validation and test procedures.

In addition, the proposed solution is highly scalable. The scheme applies to any speed. Its application is only limited by the performance of the ATE instrument. Higher speed applications (such as 10Gbps) are attainable in the future with the advance in ATE instruments.

Furthermore, we have presented the infrastructure that can be used standalone or in conjunctions with an ATE, with the mean objective of being "ahead of the curve" in high-speed serial interface testing. In other words, when ATE equipment is not available to test at a newly achieved speed of the manufactured devices, our infrastructure, based on discrete relays (including MEMS-based ones), can come to the rescue.

The proposed ATE approaches have been successfully used in production to test millions of devices. They also have been used for the validation of new designs to quickly find design issues and search for the worst case devices across PVT corners. With the approaches, we have provided a lot of invaluable characterization data within very short time. All these are impossible to achieve by just relying on bench work.

As another aspect of our work, we have made the ATE play a more and more important role in more complete validation and characterization besides its traditional role in less involved production testing.

We also propose non-ATE based HSSI jitter qualification solutions. Using the state-of-the-art phase delay lines, we can inject controllable amounts of jitter to test HSSIs with data rates up to 12.5 Gbps. The low cost solution not only eliminates the need for expensive high-speed ATE instruments, it also meets the test need for applications with data rates above 6Gbps where no ATE instruments are available. In addition, we also present FPGA-based BERTs that can be used to check the bit errors and verify the functionality of HSSIs.

Finally, we also investigate the amplitude noise impact to BER. We present a novel scheme to implement the digital AWGN generator that is suitable for low BER evaluation. The proposed generator, and the complete BERT infrastructure around it, has a number of unique characteristics, not matched by any similar design. For instance, our AWGN generator can achieve the largest values achievable by any (zero-mean, unity-value) Gaussian generator, at a very modest cost in hardware.

Reference

[1] Y. Fan and Z. Zilic, "Qualifying Serial Interface Jitter Rapidly and Cost-effectively," Springer Journal of Electronic Testing: Theory and Applications, Volume 26, 2010, 17 pages, DOI: 10.1007/s10836-009-5131-5

[2] Y. Fan, Y. Cai, L. Fang, A. Verma, B. Burcanowski, Z. Zilic and S. Kumar, “An Accelerated Jitter Tolerance Test Technique on ATE for 1.5GG/s and 3GB/s Serial-ATA,” Proceedings of IEEE International Test Conference, Oct. 2006

[3] Y. Fan and Z. Zilic "Accelerating Jitter Tolerance Qualification for High Speed Serial Interfaces," Proceedings of the 10th International Symposium on Quality Electronic Design, ISQED'09, March. 2009

[4] Y. Fan, Y. Cai and Z. Zilic, "A High Accuracy, High Throughput Jitter Test Solution on ATE for 3 Gbps and 6 Gbps Serial-ATA," Proceedings of IEEE International Test Conference, Oct. 2007

[5] Y. Fan and Z. Zilic, "A Versatile Scheme for Validation, Testing and Debugging of High Speed Serail Interfaces," Proceedings of IEEE High Level Design Validation and Test Workshop, HLDVT’09, Nov. 2009

[6] Y. Fan and Z. Zilic, “Bit Error Rate Testing of Communication Interfaces,” IEEE Transactions on Instrumentation and Measurements, Vol. 57, No. 5, pp. 897-906, May 2008

[7] P. Patra, “On the Cusp of a Validation Wall,” IEEE Design & Test, volume 24, Issue 2, PP.193-196, Mar.-Apr. 2007

[8] ITRS. The International Technology Roadmap for Semiconductors, 2007 Edition

[9] IEEE 802.3 Ethernet Working Group. http://www.ieee802.org/3/

[10] Altera Corporation. Mercury Programmable Logic Device Family Data Sheet, San Jose, California, January 2003

[11] Altera Corporation. Stratix IV Device Datasheet, December 2008 http://www.altera.com/literature/hb/stratix-iv/stx4_5v4_01.pdf

[12] Xilinx, Inc. Introducing Virtex-6 and Spartan-6 FPGA Families, http://www.xilinx.com/products/v6s6.htm

[13] Texas Instruments Incorporated http://www.ti.com

[14] Cisco Systems Inc. http://www.cisco.com

[15] P. Noel, F. Zarkeshvari and T. Kwasniewski, “Recent Advances in High-Speed Serial I/O Trends, Standards and Techniques,” Proceedings of 18th Canadian Conference on Electrical and Computer Engineering, 2005

[16] H. Johnson and M. Graham, High-Speed Signal Propagation Advanced Black Magic, Prentice Hall PTR, 2003

[17] A. Hajimiri and T. H. Lee, The Design of Low Noise Oscillators, Kluwer Academic Publishers, 1999

[18] T. Miyazaki, M. Hashimoto and H. Onodera, “A Performance Prediction of Clock Generation PLLs: A Ring Oscillator Based PLL and an LC Oscillator Based PLL,” IEICE Transactions on Electronics 2005 E88-C (3): 437-444

Y. Fan, Z. Zilic, *Accelerating Test, Validation and Debug of High Speed Serial Interfaces*,
DOI 10.1007/978-90-481-9398-1,

[19] Altera Corporation, Innovating with a Full Spectrum of 40-nm FPGAs and ASICs with Transceivers, white paper

[20] M. P. Li, Jitter, Noise, and Signal Integrity at High-Speed, Prentice Hall, 2007

[21] Altera Corporation, The Evolution of High-Speed Transceiver Technology, White Paper, San Jose, California, 2002

[22] Serial ATA International Organization: Serial ATA Revision 3.0. Gold Revision, June 2, 2009

[23] National Committee for Information Technology Standardization (NCITS) T11.2/Project 1316-DT/Rev 3.1: "Fiber Channel – Methodologies for Jitter and Signal Quality Specification", October 2001

[24] IEEE Draft P802.3ae/D3.3, "Supplement to Carrier Sense Multiple Access with Collision Detection (CSMA/CD) Access Method & Physical Layer Specifications," XGMII Extended Sublayer (XGXS) and 10 Gigabit Attachment Unit Interface (XAUI), October 2001

[25] SATA-IO, Serial ATA International Organization. http://www.sata-io.org/

[26] "Serial ATA: Meeting Storage Needs Today and Tomorrow - Introducing SATA 6Gb/s and the Serial ATA Revision 3.0 Specification," SATA-IO SATA Rev 3.0 Presentation, May 27-June 6, 2009. http://www.serialata.org/documents/SATA-Rev-30-Presentation.pdf

[27] "Economic Crisis Response: Worldwide 2008-2012 Forecast Update," IDC Doc #215614, December 2008

[28] E. A Newcombe and S. Pasupathy, "Error Rate Monitoring for Digital Communications," Proceedings of the IEEE, volume 70, Issue: 8, Aug.1982, pp.805-828

[29] D.H. Wolaver, "Measure Error Rate Quickly and Accurately," Electronic Design, pp.89-98, May 30, 1995

[30] S. Berber, "An Automated Method for BER Characteristics Measurement," IEEE Transactions on Instrumentation and Measurement, VOL. 53, No. 2, April 2004

[31] A. Papoulis, Probability, Random Variables, and Stochastic Processes, New York: McGraw-Hill, 1984

[32] K. S. Shanmugan and A.M. Breipohl, Random Signals: Detection, Estimation, and Data Analysis, New York, John Wiley and Sons, 1988

[33] Maxim Integrated Products, Inc. HFTA-05.0: Statistical Confidence Levels for Estimating Error Probability, Application Notes, 2007

[34] Y. Cai, S. Werner, G. Zhang, M. Olsen, and R. Brink, "Jitter Testing for Multi-gigabit Backplane SerDes – Techniques to Decompose and Combine Various Types of Jitter," Proceedings of IEEE International Test Conference, 2002, p700-709

[35] John Patrin and Mike Li, "Comparison and Correlation of Signal Integrity Measurement Techniques," DesignCon 2002

[36] "Jitter Fundamentals," Wavecrest Corporation, Rev.1. July 6, 2005. http://www.wavecrest.com/technical/pdf/jittfun_hires_sngls.pdf

[37] A. Kuo, T. Farahmand, N. Ou, S. Tabatabaei, and A. Ivanov, "Jitter Models and Measurement Methods for High-Speed Serial Interconnects," Proceedings of IEEE International Test Conference, 2004

[38] J. G. Proakis, Digital Communications, McGraw-Hill High Education, 2001
[39] http://www.altera.com/products/devices/mercury/features/mcy-cdr.html "Mecury Advanced CDR Support", Altera Application Notes
[40] B. Laquai and Y. Cai "Testing Multilane Gigabit SerDes Interfaces with Jitter Injection," Proceedings of IEEE International Test Conference, 2001
[41] M. Hafed, D. Watkins, C. Tam, and B. Pishdad, "Massively Parallel Validation of High-speed Serial Interfaces Using Compact Instrument Modules," Proceedings of IEEE International Test Conference, 2006
[42] D.C. Keezer, D. Minier, and P. Ducharme, "Source-Synchronous Testing of Multilane PCI Express and HyperTransport Buses," IEEE Design & Test of Computers, January 2006
[43] http://www.advantest.de/dasat/index.php?cid=100354&conid=100984&
[44] S. Sunter and A. Roy, "Structural Tests for Jitter Tolerance in SerDes Receivers," Proceedings of IEEE International Test Conference, 2005
[45] M. S. Toden, Analog and Digital Communication Systems, Discovery Press, Los Angeles, California, 2001
[46] The MathWorks, Inc. Natick, Massachusetts. http://www.mathworks.com
[47] A. Gazel, E. Boutillon, J.L. Danger, G. Gulak, and H. Lamaari, "Design and Performance Analysis of a High Speed AWGN Communication Channel Emulator," Proceedings of IEEE PACRIM Conference, Aug. 2001
[48] Altera Corporation, San Jose, California, http://www.altera.com
[49] Xilinx, Inc. San Jose, California, http://www.xilinx.com
[50] M. Courtoy, "Rapid System Prototyping for Real-time Design Validation," Proceedings of Ninth International Workshop on Rapid System Prototyping, 1998, pp. 108-112
[51] T. Matsumura, N. Yamanaka, T. Yamaguchi, and K. Ishikawa, "Real-time emulation Method for ATM Switching Systems in Broadband ISDN," Proceedings of seventh IEEE International Workshop, Jun 1996
[52] Wai-Kei Mak and D. F. Wong, "Board-level Multiterminal Net Routing for FPGA-based Logic Emulation," ACM Transactions on Design Automation of Electronic Systems (TODAES), April 1997
[53] J. Chen, J. Moon and K. Bazargan, "A Reconfigurable FPGA-Based Readback Signal Generator for Hard-Drive Read Channel Simulator," Proceedings of ACM/IEEE Design Automation Conference, June 10-14, 2002
[54] C. Change, K. Kuusilinna, B. Richards and R.W. Brodersen, "Implementation of BEE: a Real-time Large-scale Hardware Emulation Engine," Proceedings of the International Symposium on Field programmable Gate Arrays, February 2003
[55] Agilent Technologies, Agilent 81200 Data Generator/Analyzer Data sheet, 2002
[56] Anritsu Corporation, 48 Gb/s BER Test System Datasheet. Atsugi-shi, Kanagawa, Japan, 2002
[57] M. F. Schollmeyer and W. H. Tranter, "Noise Generators for the Simulation of Digital Communication Systems," Proceedings of the 24th Annual Simulation Symposium, April 1-5, 1991, pp. 264 – 275
[58] P. Kabal, "Generating Gaussian Pseudo-random Deviates", Tech. Rep., Department of Electrical and Computer Engineering, McGill University, 2000

[59] D. B. Thomas, W. Luk, P. Leong, and J. D. Villasenor, "Gaussian Random Number Generators," ACM Computing Surveys, Vol. 39, No. 4, October 2007

[60] Xilinx Inc., "Additive White Gaussian Noise (AWGN) Core v1.0", Production Specification, Oct. 2002

[61] D. Lee, W. Luk, J. Villasenor, and P. Y. K. Cheung, "A Gaussian Noise Generator for Hardware-Based Simulators," IEEE Transactions on Computers, Vol. 53, No. 12, December 2004

[62] A. Alimohammad, B.F. Cockburn, and C. Schlegel, "An iterative hardware Gaussian noise Generator," Proceedings of IEEE Pacific Rim Conference on Communications, Computers and Signal Processing, Aug. 2005

[63] A. Alimohammad, S.F. Fard, B.F. Cockburn, and C. Schlegel, "A compact and accurate Gaussian Variate Generator," IEEE Transactions on Very Large Scale Integration (VLSI) Systems, Vol.16, No.5, May 2008

[64] P. Atiniramit, Design and implementation of an FPGA-based adaptive filter single-use receiver, M. Sc. Thesis, Virginia Polytechnic Institute, 1999

[65] D. E. Knuth, The Art of Computer Programming, Addison-Wesley, 1998

[66] S. Aouini and G. W. Roberts, "A Predictable Robust Fully Programmable Analog Gaussian Noise Source for Mixed-Signal/Digital ATE," Proceedings of IEEE International Test Conference, 2006

[67] Y. Fan and Z. Zilic, "A Novel Scheme of Implementing High Speed AWGN Communication Channel Emulators in FPGAs," Proceedings of 2004 International Symposium on Circuits and Systems, Volume: 2, 23-26, May 2004

[68] Henry H. Y. Chan and Zeljko Zilic, "Modeling Simultaneous Switching Noise-Induced Jitter for System-on-Chip Phase-Locked Loops," Proceedings of ACM/IEEE Design Automation Conference, June 2007

[69] Y. Cai, B. Laquai, and K. Luehman, "Jitter Testing for Gigabit Serial Communication Transceivers," IEEE Design & Test of Computers, Vol. 19, Issue 1, Jan, 2002

[70] "Fibre Channel - Methodologies for Jitter and Signal Qualify Specification - MJSQ Technical Report, Revision 12.2," Ed. Bill Hamn, January 2004

[71] M. Li, "Statistical and System Approaches for Jitter, Noise, and Bit Error Rate (BER) Tests for High Speed Serial Links and Devices," Proceedings of IEEE International Test Conference, 2005

[72] S. Sunter and A. Roy, "A Self-Testing BOST for High-Frequency PLLs, DLLs and SerDes," Proceedings of IEEE International Test Conference, 2007

[73] R. Walker, "Designing Bang-bang PLLs for Clock and Data Recovery in Serial Data Transmission Systems," http://www.omnisterra.com/walker/pdfs.papers/BBPLL.pdf White Paper, 2003

[74] Behzad Razavi, "Challenges in the Design of High-Speed Clock and Data Recovery Circuits," IEEE Communication Magazine, August 2002

[75] D. Hong and K.T. Cheng, "Bit-Error Rate Estimation for Bang Bang Clock and Data Recovery Circuitry in High-Speed Serial Links," Proceedings of 26th IEEE VLSI Test Symposium, 2008

[76] Working Draft American National Standard, Serial Attached SCSI-2 (SAS-2), Revision 5a, 21 July 2006

[77] T. Palkert, "SFI-5 Proposed Electrical Specifications," Optical Internetworking Forum (OIF2001.033), January 2001
[78] "High Frequency Serial Communication: Technology Requirement", Test and Test Equipment Section, ITRS: International Technology Roadmap for Semiconductors, Nov, 2004
[79] D.C. Keezer, J.S. Davis, S. Bezos, D. Minier, M.C. Caron, K. Bergman, and O. Liboiron-Ladouceur, "Low-Cost Strategies for Testing Multi-Gigahertz SOPs and Components," Proceedings of IEEE 5th Electronics Packaging Technology Conference, 2003
[80] M. Li and J. B. Wilstrup, "On the Accuracy of Jitter Separation from Bit Error Rate Function," Proceedings of IEEE International Test Conference, 2002, p710-716
[81] P. R. Trischitta and E. L. Varma, Jitter in Digital Transmission Systems, Artech House, 1989
[82] Bellcore, SONET OC-192 Transport System Generic Criteria, GR-1377-CORE, Issue 5, 1998
[83] T.J. Yamaguchi, M. Soma, M. Ishida, H. Musha, and L. Malarsie, "A New Method for Testing Jitter Tolerance of SerDes Devices Using Sinusoidal Jitter," Proceedings of IEEE International Test Conference, 2002
[84] Teradyne, Inc. http://www.teradyne.com
[85] H. Werkmann, "Enabling the PCI Express Ramp - ATE Based Testing of PCI Express Architecture," Euro DesignCon 2004 Also available at www.verigy.com
[86] Y. Cai, A. Bhattacharyya, J. Martone, A. Verma, and W. Burchanowski, "A Comprehensive Production Test Solution for 1.5GB/S and 3GB/S Serial-ATA," Proceedings of IEEE International Test Conference, 2005
[87] Y. Cai, T. P. Warwick, S. G. Rane, and E. Masserrat, "Digital Serial Communication Device Testing and Its Implications on Automatic Test Equipment Architecture," Proceedings of IEEE International Test Conference, 2000
[88] E.S. Erdogan and S. Ozev, "A Robust, Self-tuning CMOS Circuit for Built-in Go/No-Go Testing of Synthesizer Phase Noise," Proceedings of IEEE International Test Conference, 2006
[89] E.S. Erdogan and S. Ozev, "An ADC-BiST Scheme Using Sequential Code Analysis," Proceedings of the conference on Design, Automation and Testing in Europe, 2007
[90] M. Li and J. Chen, "New Methods for Receiver Internal Jitter Measurements," Proceedings of IEEE International Test Conference, 2007
[91] Tektronix, "Jitter Measurement and Timing Analysis," Product guideline, http://www.tek.com/applications/serial_data/jitter.html
[92] Agilent Technologies, "Jitter Solutions for Telecom, Enterprise, and Digital Designs," Cooperate literature, http://cp.literature.agilent.com/litweb/pdf/5988-9592EN.pdf
[93] B.A. Ward, K. Tan, and M.L. Guenther, "Apparatus and Method for Spectrum Analysis-Based Serial Data Jitter Measurement," US Patent No. 6832172, issued on December 14, 2004

[94] W. Dalal and D. Rosenthal, “Measuring Jitter of High Speed Data Channels Using Undersampling Techniques,” Proceedings of IEEE International Test Conference, 1998

[95] M. Li, J.Wilstrup, R. Ressen and D. Petrich, “A New Method for Jitter Decomposition through Its Distribution Tail Fitting,” Proceedings of IEEE International Test Conference, 1999

[96] S. Sunter, A. Roy, and J. Cote, “An Automated, Complete, Structural Test Solution for SERDES,” Proceedings of IEEE International Test Conference, 2004

[97] A. H. Chan and G. W. Roberts, “A Jitter Characterization System Using a Component-Invariant Vernier Delay Line,” IEEE Transactions on VLSI Systems, Volume 12, Issue 1, Jan. 2004

[98] A. Meixner, A. Kakizawa, B. Provost, and S. Bedwani, “External Loopback Testing Experience with High Speed Serial Interfaces,” Proceedings of IEEE International Test Conference, Oct. 2008

[99] W. Fritzsche and A. Haque, “Low Cost Testing of Multi-GBit Device Pins with ATE Assisted Loopback Instrument,” Proceedings of IEEE International Test Conference, Oct. 2008

[100] G. Hansel, K. Stieglbauer, K. Schulze and J. Moreira, “Implementation of an Economic Jitter Compliance Test for a Multi-Gigabit Device on ATE”, Proceedings of IEEE International Test Conference, 2004

[101] Standard Error (Statistics) http://en.wikipedia.org/wiki/Standard_error_(statistics)

[102] “Standard Error of the Mean,” material from Brighton Webs Ltd, statistical and data services for industry. http://www.brighton-webs.co.uk/statistics/standard_error.asp

[103] M. Burns and G. W. Roberts, An Introduction to Mixed-Signal IC Test and Measurement, Oxford University Press, 2001

[104] Analyzing Jitter Using a Spectrum Approach, Tektronix Application note

[105] J. Dorsch, "The Softer Side of Test: Software Products to Star at International Test Conference," Electronic News, September 1999

[106] S. Sunter, C. McDonald and G. Danialy, "Contactless Digital Testing of IC Pin Leakage Currents," Proceedings of International Test Conference, 2001

[107] S. Sunter and A. Roy, "Purely Digital BIST for Any PLL or DLL," Proceedings of European Test Symposium, May 2007

[108] R. Kiefer, Test Solutions for Digital Networks, Huthig GmbH, Heidelberg, 1998

[109] S. Brown and Z. Vranesic, Fundamentals of Digital Logic with VHDL Design, McGraw-Hill Higher Education, 2000

[110] Altera Corporation, Mercury Gigabit Transceiver MegaCore Function User Guide, February 2002

[111] T. Yamaguchi, “Loopback or Not,” Proceedings of IEEE International Test Conference, 2004, p. 1434

[112] http://en.wikipedia.org/wiki/Delay_line

[113] http://www.maxim-ic.com/appnotes.cfm/an_pk/209, “How Delay Lines Work,” Application notes

[114] K.W. Kobayashi, "A DC-40 GHz InP HBT Gilbert Multiplier," Proceedings of IEEE Gallium Arsenide Integrated Circuit (GaAs IC) Symposium, 25th Annual Technical Digest 2003
[115] http://www.gigoptix.com/pdf/whitepapers/Optical_Interconnection_rev6.pdf, " Semiconductor Technologies for Optical Interconnection, from Ultra Long Haul to Very Short Reach," white paper by GigOptix
[116] www.hikari-trading.com/opt/gigoptix/file/it4036/it4036_data.pdf, "iT4036 Wideband Phase Delay Preliminary Datasheet"
[117] J. Moreira, H. Barnes and G. Hoersch, "Analyzing and Addressing the Impact of Test Fixture Relays for Multi-Gigabit ATE I/O Characterization Application," Proceedings of IEEE International Test Conference, 2007
[118] D.C. Keezer, D. Minier, P. Ducharme, D. Viens, G. Flynn, and J.S. McKillop, "Multi-GHz Loopback Testing Using MEMS Switches and SiGe Logic," Proceedings of IEEE International Test Conference, October 2007
[119] "TT1244: SPDT 265.GHz RF MEMS Switch", Data Sheet, TeraVicta, http://www.rapidtek.net/spec/mems/DS-TT1244_1.3.pdf
[120] "TeraVicta Announces Availability of DC to 26.5 GHz SPDT MEMS Switch," Press Release, TeraVicta Technologies, Inc., Austin, Texas, April 30, 2007. http://news.thomasnet.com/fullstory/518285
[121] http://www.teledynerelays.com/pdf/electromechanical/grf300grf303.pdf, GRF300 Series data sheet
[122] http://www.teledynerelays.com/pdf/SurfacemountingGRF300.pdf, "Surface Mounting GRF300 and GRF303 relays", Application Note
[123] R. D'Agostino and M. Stephens, Goodness-of-Fit Techniques, Marcel Dekker Inc., 1986
[124] D. Lee, W. Luk, J. Villasenor, G. Zhang and P. Leong, "A Hardware Gaussian Noise Generator Using the Wallace Method," IEEE Transactions on Very Large Scale Integration (VLSI) Systems, Vol. 13, No. 8, August 2005
[125] D. Knuth, The Art of Computer Programming, Seminumerical Algorithms, 3rd Edition Ser. Addison-Wesley, 1997 Vol. 2
[126]D. Ruppert, "What is Kurtosis? An Influence Function Approach," The American Statistician, Vol. 41, No. 1, pp. 1-5, Feb. 1987
[127]D.N. Joanes and C.A. Gill, "Comparing Measures of Sample Skewness and Kurtosis," Journal of the Royal Statistical Society (Series D): The Statistician 47 (1), 183–189, 1998
[128] http://en.wikipedia.org/wiki/Kurtosis
[129] D. Lee, J. Villasenor, W. Luk and P. Leong, "A Hardware Gaussian Noise Generator Using the Box-Muller Method and Its Error Analysis," IEEE Transactions on Computers, 2006
[130] D. C. Keezer, D. Minier and P. Ducharme, "Method for Reducing Jitter in Multi-Gigahertz ATE," Proceedings of the Conference on Design, Automation and Test in Europe, 2007, Pages 701-706
[131] D. C. Keezer, C. Gray, A. Majid and P. Ducharme, "A Development Platform and Electronic Modules for Automated Test Up to 20Gbps," Proceedings of IEEE International Test Conference, 2009

[132] K-L. Lim and Z. Zilic, "An Undersampled Duty Cycle Jitter BIST Circuit", Proceedings of IEEE Midwest/NEWCAS Symposium, pp. 201-204, Aug. 2007.

[133] M. Hafed, N. Abaskharoun and G. W. Roberts, "A Stand-Alone Integrated Test Core for Time and Frequency Domain Measurements," Proceedings of IEEE International Test Conference, 2000

[134] IEEE 1149.6 http://grouper.ieee.org/groups/1149/6/

[135] J.S. Davis, D.C. Keezer, O. Liboiron-Ladouceur and K. Bergman, "Application and Demonstration of a Digital Test Core: Optoelectronic Test Bed and Wafer-level Prober," Proceedings of IEEE of the International Test Conference, 2003

[136] IEEE 1149 familiy. www.siliconaid.com/2006_SWDFT_presentations/IEEE%201149%20family%20%5BRead-Only%5D.pdf

[137] D.C. Keezer, J.S. Davis, M. Haris, S. Bezos, D. Minier, M.C. Caron, K. Bergman, and O. Liboiron-Ladouceur, "Recent Advances in Low-Cost Multi-GigaHertz Testing," Proceedings of Napa KDG Packaging and Test Workshop, 2003

[138] C. Gray, O. Liboiron-Ladouceur, D.C. Keezer, and K. Bergman, "Co-Development of Test Electronics and PCI Express Interface for a Multi-Gbps Optical Switching Network," Proceedings of IEEE International Test Conference, 2007

[139] B. R. Veillette and G. Roberts, "Reliable Analog Bandpass Signal Generation," Proceedings of IEEE International Symposium on Circuits and Systems, 1998

[140] IEC 61280-2-8, "Fibre Optic Communication Subsystem Test Procedures – Digital Systems – Part 2-8: Determination of low BER using Q-factor Measurements," Feb 2003

[141] K. Willox, "Q Factor: The Wrong Answer for Service Providers and Equipment Manufactures," IEEE Communications Magazine, Feb. 2003

[142] GigOptix, http://www.gigoptix.com.

[143] P. Landman, "A Transmit Architecture with 4-Tap Feedforward Equalization for 6.25/12.5Gb/s Serial Backplane communications", Proceedings of IEEE International Solid-State Circuits Conference, 2005

[144] J. Li and F. Yuan, "A New Hybrid Phase Detector for Reduced Lock Time and Timing Jitter of Phase Locked Loops", Journal of Analog Integrated Circuits and Signal Processing, Vol. 56, Issue 3 (September 2008), Pages: 233-240

[145] S. Mehrmanesh and N. Masoumi, "A Comprehensive Bang-Bang Phase Detector Model for High Speed Clock and Data Recovery Systems", Proceedings of the 17th International Conference on Microelectronics, 2005. ICM 2005

[146] J. D. H. Alexander, "Clock Recovery from Random Binary Signals", Electronics Letters, Vol. 11, Issue 22 (Octomber 1975), Pages: 541-542

[147] J. Lee, K. S. Kundert and B. Razavi, "Analysis and Modeling of Bang-Bang Clock and Data Recovery Circuits", IEEE Journal of Solid-State Circuits, Vol. 39, Issue 9 (September 2004), Pages: 1571-1580

[148] MSS (Mobile Satellite Services) Corporation, Bit Error Rate Generator and Additive White Gaussian Noise Generator Specification, Gaithersburg, Maryland

[149] VR. C. Tausworthe, "Random Numbers Generated by Linear Recurrence Modulo Two," Mathematical Computing, vol. 19, pp201-209, 1965

[150] S. W. Golomb, Shift Register Sequences, Holden-Day, San Francisco, 1967

[151] S. Wolfram, "Random Sequence Generation by Cellular Automata," Advances in Applied Mathematics, vol. 7, pp. 123-169, June 1986

[152] J. Chen, J. Moon and K. Bazargan, "A Reconfigurable FPGA-based Readback Signal Generator for Hard-drive Read Channel Simulator," Proceedings of 39th Design Automation Conference, Pages: 349-354, June 2002

[153] S. Wolfram, "Statistical Mechanics of Cellular Automata," Reviews of Modern Physics, vol. 55, pp.601-644, July 1983

[154] S. Zhang, D. M. Miller, J. C. Muzio, "Determination of Minimal Cost One-dimensional Linear Hybrid Cellular Automata," Electronics Letters, vol. 27, no.18, pp.1625-1627, Aug. 1991

[155] K. Cattell, S. Zhang, "Minimal Cost One-dimensional Linear Hybrid Cellular Automata of Degree through 500," Journal of Electronic Testing: Theory & Applications, vol6, no.2, pp255-258, April 1995

[156] M. Sipper and M. Tomassini, "Generating Parallel Random Number Generators by Cellular Programming," International Journal of Modern Physics C, vol. 7, no. 2, pp.181-190, 1996

[157] M. Tomassini, M. Sipper, M. Zolla, and M. Perrenoud, "Generating High-quality Random Numbers in Parallel by Cellular Automata," Future Generation Computer Systems, vol. 16, pp. 291-305, 1999

[158] M. Tomassini, M. Sipper, and M. Perrenoud, "On the Generation of High-quality Random Numbers by Two-dimensional Cellular Automata," IEEE Transactions on Computers, vol.49, pp.1146-1151, October 2000

[159] B. Shackleford, M. Tanaka, R. J. Carter and G. Snider, "FPGA Implementation of Neighbourhood-of-four Cellular Automata Random Number Generators," Proceedings of the Tenth ACM International Symposium on Field-Programmable Gate Arrays, pp.106-112, Feb. 2002

[160] F. James, "A Review of Pseudo-random Number Generators," Computer Physics Communications 60, 1990

[161] P. L`Ecuyer, "Random Numbers for Simulation," Communications of the ACM, 33:10, 1990

[162] G.A. Marsaglia, "A Current View of Random Number Generators," Computational Science and Statistics: The Interface, Balliard, Elsevier, Amsterdam, 1985

[163] Quantum World Corporation, QNG Model J20KP True Random Number Generator Users Manual, 1998

[164] P. H. Bardell, W.H. McAnney and J. Savir, Build-in Test for VLSI: Pseudo-random Techniques, John Wiley and Sons, 1987

[165] P.Alfke, "Efficeent Shift Registers, LFSR Counters, and Long Psudo-Random Sequence Generators," Xilinx Application Note, 1995

[166] P. Chu and R. Jones, "Design Techniques of FPGA-Based Random Number Generator," Military and Aerospace Applications of Programmable Devices and Technologies Conferences, 1999

[167] P. L`Ecuyer, "Efficient and Portable Combined Random Number Generators," Comm. ACM 31:6, 1988

Index

FSC
www.fsc.org
MIX
Papier | Fördert
gute Waldnutzung
FSC® C083411

Kolibri 360 GmbH
Ferdinand-Jühlke-Straße 7
99095 Erfurt, Deutschland
produktsicherheit@kolibri360.de